Sustainable Structures and Buildings

Alireza Bahrami
Editor

Sustainable Structures and Buildings

Springer

Editor
Alireza Bahrami
Department of Building Engineering, Energy Systems
and Sustainability Science, Faculty of Engineering
and Sustainable Development
University of Gävle
Gävle, Sweden

ISBN 978-3-031-46687-8 ISBN 978-3-031-46688-5 (eBook)
https://doi.org/10.1007/978-3-031-46688-5

This work was supported by University of Gävle, Sweden

This Springer imprint is published by the registered company Springer Nature Switzerland AG
The registered company address is: Gewerbestrasse 11, 6330 Cham, Switzerland

Paper in this product is recyclable.

This book is dedicated to my unforgettable dear father

Preface

This book encourages monitoring methods and advancements required for resilient structures, addressing the most crucial problems for creating sustainable and cutting-edge solutions for the construction and building industry. The book also covers how civil engineers may improve the design of structures for more environmentally friendly construction, buildings, and building materials in order to assist in achieving the sustainable development goals (SDGs) of the United Nations (UN). With a particular emphasis on the sustainable solutions, it provides strategies and plans for developing top-notch recommendations in civil engineering.

The importance of different structures, buildings, and building materials in sustainable building and construction brings together a recognized group of individuals from diverse cultures and building traditions to discuss why it is time to reevaluate our construction methods and think about replacing many of the low-performance materials that are currently used in the industry. This book enhances the theory and practice of sustainable building design by providing important practical tools that might influence building practices in both the public and private sectors. It additionally encourages the ongoing efforts to establishing and executing SDGs.

As a more sustainable alternative to steel and concrete, cross-laminated timber techniques are increasingly utilized in mid-rise and high-rise structures. However, this raises serious concerns about the fire performance of these structures. It is discovered that many innovative structures produce deformation modes that are predictable, ductile, and without of timber damage. Structural textiles and foils, a type of lightweight materials, have recently gained recommendations among designers and architects thanks to their attractive qualities. Bamboo has the potential to be used as a tensile component in concrete. Currently, the handling of mine waste is still based on a linear approach, which needs to be taken into account for its storage and landfilling. Application of mine waste as a new raw material can be a step toward the circular economy. Mine waste can contribute to the reduction of natural resource consumption and CO_2 emissions. A change from fired clay bricks and cement bricks to geopolymer bricks would remarkably help solve the existing housing and agricultural biomass waste problems. Reinforced concrete constructions automatically experience micro-cracks development due to concrete shrinkage, freezing, and thawing. Bacterial concrete is an effective approach for repairing concrete and enhancing its longevity. Urban drainage infrastructures provide a number of essential services in urban areas and are necessary to meeting SDGs to address current problems and build a more sustainable future.

All the aforementioned subjects are thoroughly explored and presented in this book. This book has been written for engineering students, researchers, and professionals who are interested in fostering and implementing sustainability at distinct phases of planning, constructing, and maintaining structural projects and buildings.

Gävle, Sweden
2024

Alireza Bahrami

Acknowledgements

The authors are thankful to the strategic research area Urban Sustainability at the University of Gävle, Sweden for funding this book.

Contents

Chapter 1
Sustainable Development of Recent High-Rise Timber Buildings

Alireza Bahrami and S. M. Priok Rashid

1.1 Introduction

High-rise buildings may be designed and constructed with increasing complexity because of technological advancements. Buildings that are over 25 m can be categorized as "high-rise buildings" [1]. It can be assumed that taller buildings have bigger areas, more premises, higher energy consumptions, and more CO_2 emissions, thus, creating a bigger negative impact on the environment. High-rise building construction is often expensive, takes a long time, and requires a lot of labor. In contrast, high-rise buildings are distinctive because of the direct influence they have on the surrounding environment due to their height and architectural design.

Börjesson and Gustavsson [2] revealed that the primary energy input (mainly fossil fuels) in the production of building materials was about 60–80% lower for timber frames compared with concrete frames. According to Oliver et al. [3], using wood substitutes could save 14–31% of global CO_2 emissions and 12–19% of global fossil fuel consumption by utilizing 34–100% of the world's sustainable wood growth.

Till now, reinforced concrete, steel, and glass have been the most popular materials utilized to build high-rise buildings for historical reasons. In the late nineteenth century, towns were frequently affected by fires, resulting in the adoption of fire safety measures in numerous countries throughout the world, including laws that restrict the use of timber structures in the construction of high-rise buildings. Recently, the fast development of construction technology, the introduction of

A. Bahrami (✉)
Department of Building Engineering, Energy Systems and Sustainability Science,
Faculty of Engineering and Sustainable Development, University of Gävle, Gävle, Sweden
e-mail: Alireza.Bahrami@hig.se

S. M. P. Rashid
Department of Civil Engineering, Faculty of Engineering, Hajee Mohammad Danesh Science and Technology University, Dinajpur, Bangladesh

A. Bahrami (ed.), *Sustainable Structures and Buildings*,
https://doi.org/10.1007/978-3-031-46688-5_1

modern engineered timber products (such as cross-laminated timber (CLT) and glued laminated timber (glulam)), and the modification of construction restrictions prompted a new era, which led to the growth of multi-story timber construction [4]. As previously mentioned, the construction of high-rise timber buildings directly contributes to the success of sustainable construction goals across the three aspects of sustainability.

The implementation of CLT in Australia has been gradual for a variety of factors, including cultural, structural, economic, and maintenance issues [5]. Perhaps the most important is the traditional cultural resistance to change in construction and a misconception by industry professionals and users that engineered timber possesses the same traditional weaknesses associated with timber, i.e., a susceptibility to fire and pest damage, poor acoustics, cost premiums, and issues of durability. The authors also stated that the "people" part of marketing is the reason for CLT's poor adoption, highlighting that purchasers must be encouraged to realize the advantages of a product and want it. To that purpose, goods must be developed not just to please but also to motivate buyers.

Doubts regarding the quality and traits of timber constructions are produced by a variety of perspectives in different countries. It is a frequent misconception that timber structures change in size over time, they are not stable and durable, not pest and moisture-resistant, and very flammable. Poor building procedures have formed these public impressions, where materials made of timber lost their qualities and durability owing to incorrect preparation or use.

Despite the aforementioned challenges, there has been an acceleration of interest in a new type of multi-story timber building globally. Compared with materials like steel and concrete, building materials made of timber have shown to have a significantly lower influence on climate change [6].

Recent interest in high-rise timber buildings is thanks to timber's sustainability and its other considerable advantages. Timber is the only building material that is sustainable and takes minimal energy to prepare. Timber is 500 times more thermally resistant than steel and 10 times more thermally resistant than concrete. In terms of energy efficiency, timber is a good heat insulator. By using it in construction, thermal bridges are decreased and walls' ability to retain heat is increased. Additionally, wood goods acoustically perform exceptionally well. When it comes to the transmission of impact noise, the results are better than those for concrete [7]. As they are durable, lightweight, and simple to install at a building site, timber products are an excellent option from the financial perspective [7].

Carbon is stored in all wood products. Wood is naturally made up of carbon, which trees absorb from the atmosphere as they develop. Timber products have a positive carbon impact owing to substitution and sequestration [8].

According to the 2020 Global Status Report for Buildings and Construction, annual growth in building energy usage was seen; but in 2019, energy-related CO_2 emissions rose to 9.95 $GtCO_2$. When building construction emissions were added to operating emissions, the sector was responsible for 38% of all worldwide energy-related CO_2 emissions [9].

To fulfill the Paris Agreement commitment and the sustainable development goals, the building and construction sector must reduce carbon emissions [10]. The United Nations has established legally binding guidelines for lowering greenhouse gas (GHG) emissions and implementing sustainability measures as a result of the growing threat of climate change. In this context, the European Union (EU) aims to be climate-neutral by 2050. The European Green Deal effort is focused on achieving this goal, which is also consistent with the EU's dedication to global climate action under the Paris Agreement [11].

The current chapter's objective is to investigate sustainability of modern high-rise timber buildings using multiple-criteria evaluation techniques. Modern high-rise timber buildings need to have their design evaluated more closely in order to determine their sustainability. Using wood in construction has a positive environmental impact due to several factors: (1) The only sustainable building material that takes minimal energy to prepare is wood; (2) All wood products involve carbon; (3) Wood produces less GHGs during manufacturing and does not release any CO_2 when a building is in process; and (4) Timber products are reusable and recyclable. Environmental impact studies of buildings are a crucial component of modern construction and are included in sustainable design.

1.2 Recent High-Rise Timber Buildings

Mjøsa Tower, an eighteen-story multipurpose building in Brumunddal, Norway, is now the highest timber building in the world. It was completed in March 2019 and is 85.4 m in height (Fig. 1.1). The skyscraper has a hotel, offices, and apartment buildings. The load-bearing structure is constructed of Kerto laminated veneer lumber (LVL), with glulam columns and beams. The first 10 stories are made up of prefabricated timber components, while the top story decks are made of concrete to provide weight for the building. Compared with a reinforced concrete building having the same height, prefabricated parts made it possible to finish construction far more quickly. With the majority of the materials coming from within two miles of the site, Mjøsa Tower's construction was planned to have as little impact on the environment as possible [12, 13].

At the University of British Columbia in Vancouver, Canada, Brock Commons (Fig. 1.2) is a 53 m-tall accommodation facility that can accommodate 404 students. The hybrid building of the 15,000 m^2 project consists of 17 stories of CLT floors supported by glulam columns on top of a concrete foundation and two 18-story concrete cores. The roof comprises steel beams and metal decking. The building envelope is a prefabricated panel system clad with wood-fiber high-pressure laminate. According to its designers, the building weighs 7500 tons less than a concrete counterpart [14]. Interior wood elements were coated with gypsum board to comply with fire regulations, which were tighter than those for conventional steel or reinforced concrete buildings [12, 13]. The building was constructed in 2017 using

Fig. 1.1 Mjøsa Tower, Brumunddal, Norway

Fig. 1.2 Brock Commons Tallwood House, Vancouver, Canada

Fig. 1.3 Treet, Bergen, Norway

prefabricated parts, and construction was finished 70 days after the parts were brought on-site.

In Bergen, Norway, Treet (Fig. 1.3), often known as The Tree, is a 14-story apartment block with a height of 49 m that was completed in 2015, 4 years after the design phase started in 2011. A glulam load-bearing structure, called Treet, supports 62 prefabricated modular apartments with CLT walls. The glulam structure is the only component that gives the building stiffness; however, the concrete slab floors on levels 6 and 11 add additional stability. To prevent wood on the building's façade from deteriorating, portions of it are covered in glass and metal [12, 13].

In Melbourne, Australia, the construction of a 32.2-meter-tall, nine-story building, known as Forte Development, was finished in 2012 (Fig. 1.4). It became the country's first high-rise timber building. Shear walls made of CLT panels were used to construct the building. The panels were piled at an angle, glued to the surface, and hydraulically pressed. There are four townhouses and 23 residential flats in Forte Development. The apartments are dual aspect to make the most of natural lighting, and were also designed with thermal efficiency in mind, requiring less energy to be heated [12, 13].

In Hackney, northeast London, Murray Grove Stadthaus, a nine-story residential building with a height of 26 m, completed construction in 2009. There are 29 apartments there (Fig. 1.5). Murray Grove was named the world's first housing high-rise building constructed entirely from prefabricated CLT panels, with its core, load-bearing walls, floor slabs, stairs, and lift. CLT panels were used on the building's structural core because of their noticeably higher density than timber frames. Then, independent layers were included to keep the acoustic performance good [12, 13].

Fig. 1.4 Forte Development, Melbourne, Australia

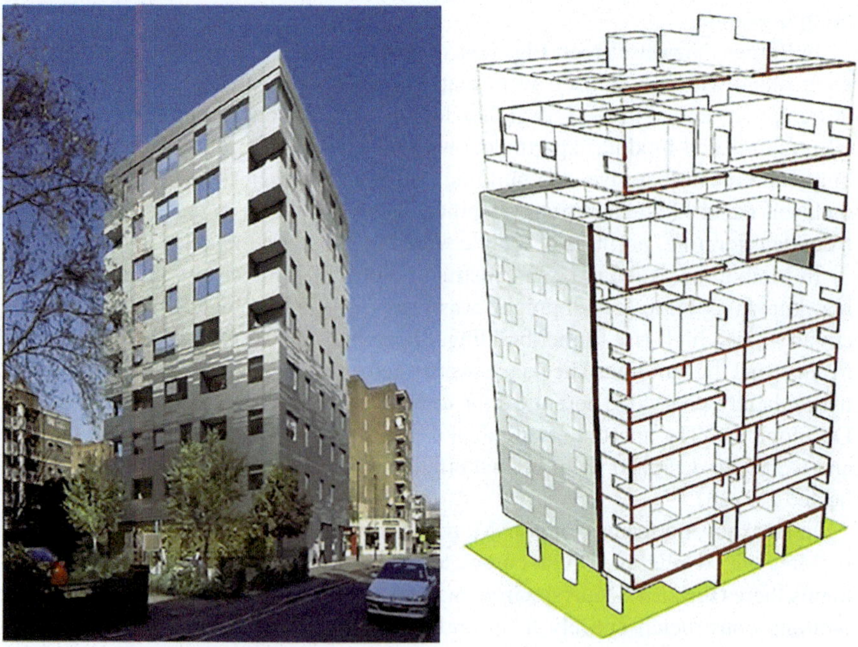

Fig. 1.5 Murray Grove Stadthaus, London, UK

Table 1.1 Trend of high-rise timber building in the world

Name of building	Location	Height (m)	Story	Completed
Treet	Bergen/NO	52	14	2014
Mjøstårnet	Brumunddal/NO	85.4	18	2019
Hoho Vienna tower	Vienna/AT	84	24	2020
Haut	Amsterdam/NL	73	21	2021
Hypérion	Bordeaux/FRA	57	18	2021
Silva	Bordeaux/FRA	50	18	2022
Ascent MKE	USA	87	25	2022

Table 1.1 displays the global trend of high-rise timber buildings. Construction is underway on further high-rise timber buildings, such as Terrace House in Vancouver, Canada.

1.3 GHGs

Buildings involve a major amount of carbon dioxide emissions, which is notable given the tremendous environmental impact of buildings' development [15]. Building materials are produced, built, used, and demolished on a large scale, which greatly adds to global resource utilization and waste. Construction and demolition activities require around 40% of all raw materials worldwide [16]. Additionally, 32% of all energy used globally and a third of all GHG emissions are related to construction activities. Buildings release 35% of GHGs, 10% of airborne particles, and 25% of landfill waste. Due to these adverse environmental impacts, sustainable alternatives such as less energy-intensive materials or recycled or biodegradable materials need to be introduced in the construction sector [17]. The circular economy of construction materials in this context also plays a noticeable role, helping maximize the efficient material use of construction and demolition waste, and minimize fossil fuel energy use and resource consumption [18]. Timber has recently been proposed by researchers as a structural material with fewer energy requirements and lower carbon emissions. Approximately 88% of all timber may be utilized to make lumber or other products with little to no waste being produced. A closed-loop circular economy is a term used to describe this kind of reuse mechanism. Construction and demolition wood or wood fiber waste is employed in engineering wood panel products, which is known as open-loop circular economy [18]. Various industrialized timber goods (such as particle board and laminated flooring) can be produced in addition to engineering wood products. Currently, 63% of waste wood is disposed of in landfills and burned without being converted into energy. Although landfilling and incinerating waste have a negative effect on environment and human health, they provide little or no energy. As waste-wood reuse, reduction, and recycling (bio-concrete, wood-plastic composite) are part of the circular economy, these activities will reduce environmental impacts [19]. For sustainable development, a systemic evaluation is thus needed to investigate the construction of wood-frame structures.

1.4 Assessment Criteria

Seven characteristics that represent the quantity of timber used in construction have been established to assess the various high-rise timber buildings. The assessment includes the achieved height in terms of meter and number of floors, duration of construction, and cost efficiency as well as environmental parameters such as reduction of CO_2 emission and energy use.

Based on the prior study and experience, the authors of the current chapter assessed the significance of the criteria in Table 1.2 [13].

Table 1.2 Assessment criteria and their significance

Criterion	Max/ Min[a]	Significance	Measuring unit	Description
Use of wood	Max	0.2	m^3/m^2	Relative indicator expresses proportion of used wooden structures (m^3) per m^2 of building. Higher use of wood is preferred.
Height of building	Max	0.2	m	Architectural height of building. The aim is to achieve tallest buildings in timber, therefore, greater height is preferred.
Number of floors	Max	0.1	Number	Number of floors in a building. Sometimes buildings of the same height may have different floors and vice versa. The higher the number of floors, the larger the area of the building can be used.
Building cost	Min	0.1	Eur/m^2	Ratio of the cost of the project to the total area. The lower the cost, the more cost-effective the project is.
Length of implementation	Min	0.1	Number of days per floor	Time in working days needed to install load-bearing structures of one floor. The shorter the installation time, the more economical the construction of the building is.
Reduction of CO_2 emissions	Max	0.2	Ton	Reduced amount of CO_2 emissions compared with a reinforced concrete building. The higher the difference, the lower the environmental impact of the timber building is.
Use of energy	Min	0.1	kW/m^2 per year	Annual energy consumption per m^2 of a building. The lower the quantity, the more energy efficient the building is, and the less negative impact it has on the natural environment.

Note: [a] Max – higher value of criterion is preferred; Min – lower value of criterion is preferred

The most sustainable building, according to the overall assessment of sustainability of buildings, is Brock Commons in Canada, while Mjøstårnet is in the second place with a slightly lower index.

These results are consistent with those of prior research by Tupenaite et al. [13], which placed Mjøstårnet as the top system across seven categories. Moreover, it can be seen that the addition of other evidence benefited in the comprehensive evaluation of the buildings, changing the rankings. In addition, values of some sustainability indicators may change over time, e.g., a new supermarket or a new kindergarten could be built, thus, accessibility of a building may increase. Updated sustainability assessments are needed if the values of indicators change.

This study confirmed the findings of Leskovar and Premrov [4], which stated that the age of timber buildings is reflected in their architecture, and the height of the buildings increases with the year of construction. Furthermore, Brock Commons and Mjøstårnet, which are higher buildings, obtained higher sustainability ratings.

Based on determined efficiency indexes, the buildings were ranked—the higher the efficiency index, the higher the rank [20]. Based on calculations, Brock Commons is the most environmentally friendly building across all sustainability rating categories. In terms of environmental indicators, this building came in second, but it outperformed other buildings with respect to social indicators. The Mjøstårnet building is in the second place, and its overall rating is not much different from that of the Brock Commons building. Brock Commons has superior socio-economic performance, while this building has the highest environmental performance. A summary assessment of sustainability of the buildings is presented in Fig. 1.6.

Fig. 1.6 Overall assessment results

1.5 Construction Cost

The bill of materials (BOMs) from the architectural designs and material pricing data from RSMeans are used to estimate the cost of building construction. RSMeans is a database of up-to-date building construction expense estimates. It includes localized industry average cost data for construction materials, labor, transportation, and storage. Plumbing design systems and energy simulation software are often able to estimate operational energy use, such as daily electricity, natural gas, and water usage.

A 12-story, 8360 m^2 mixed-use office and residential complex with CLT and glulam materials was intended as the first high-rise mass timber building in the United States. The design for the mass timber building was a detailed construction document with materials takeoff. Although the building was intended to be constructed in Portland, Oregon, the project was put on hold due to a lack of funding [21]. Then, for comparative purposes, a reinforced concrete building with the same functional fireproofing, insulation, and energy consumption results was also created. Both building designs were completed by LEVER Architecture (Portland, Oregon) with additional structural design and analysis from their partner, KPFF Engineering (Seattle). Because CLT weighs less than concrete, when utilized in buildings as structural elements, the mass timber building's overall weight was around 33% lower than that of the reinforced concrete building.

The architects and building contractors did not submit detailed cost estimates for the two planned buildings. As a result, the RSMeans 2018 database, which shows cost information for the most popular building materials, was employed to quote the material unit, labor, and overhead prices for each material listed in BOMs. For CLT, the unit cost was estimated based on the glulam cost, including the labor and overhead cost data found in RSMeans, as a proxy. These construction costs are all based on the material's regional industry average cost data in Portland, Oregon. The summarized construction costs for the two buildings, timber and reinforced concrete, grouped into building assemblies are presented in Table 1.3 and Fig. 1.7. With the data derived based on BOM supplied from the building designs and RSMeans database, the mass wood building is expected to have 88% higher total front-end costs than the reinforced concrete building in this example. This resulted from CLT and glulam's higher pricing as non-commodity goods compared with commodity items like steel and concrete. Even though manufacturers of CLT are emerging because of the increasing demand in the building sector, the price is still high as a result of limited suppliers in North America. In mass timber buildings, CLT is mainly used in the floor and wall assemblies, and glulam is utilized in the post and beam assemblies, whereas less expensive concrete is mainly employed in these assemblies for reinforced concrete buildings. If full commercialization of mass timber products is achieved, the front-end cost of high-rise mass timber buildings can become more competitive, especially with the time and cost savings during the erection of high-rise mass timber buildings compared with traditional reinforced concrete buildings [22].

Table 1.3 Two different high-rise buildings' construction costs by assembly types

	Estimation of construction cost (US$)			
Assembly type	Material	Labor	Overhead	Total
Ceiling and roof				
Mass timber building	164,819	240,260	149,794	554,873
Reinforced concrete building	98,592	126,928	80,347	305,867
Floor				
Mass timber building	2,138,568	408,303	428,117	2,974,988
Reinforced concrete building	574,012	236,041	181,824	991,877
Foundation				
Mass timber building	60,701	50,489	37,115	148,306
Reinforced concrete building	84,444	71,526	53,035	209,005
Post and beam				
Mass timber building	936,605	142,659	170,023	1,249,287
Reinforced concrete building	173,503	57,780	50,749	282,032
Wall				
Mass timber building	1,708,556	1,091,486	750,321	3,550,362
Reinforced concrete building	1,110,229	975,842	635,150	2,721,221
Total				
Mass timber building				8,477,816
Reinforced concrete building				4,510,002

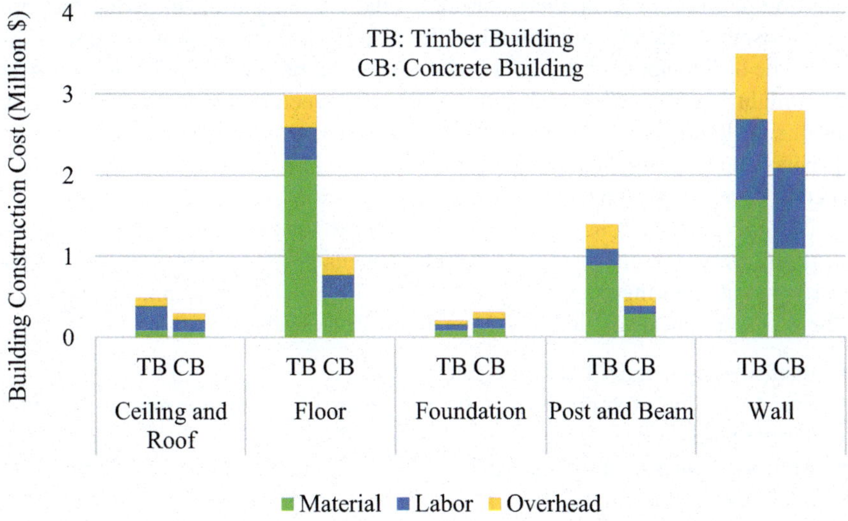

Fig. 1.7 Front-end construction costs by assembly types for high-rise mass timber and reinforced concrete (concrete) buildings

1.6 Fire Safety

High-rise timber buildings may have serious problems from a fire safety perspective. The following are the primary extra challenges for high-rise timber buildings:

- Early fire risk. Large exposed interior wood surface areas are flammable.
- A fire spread outside. Vertical fire spreads through flammable facades or cavities.
- Fires that occur during construction. This is a particular issue for structures made of light timber.
- Burnout. After fire burns itself out, what happens?
- Fire resistance which helps with the structural performance and fire containment.

Excellent fire resistance in heavy timber has been thoroughly established in the literature [23]. This exceptional behavior is due to the intense flames' gradual and predictable rate of surface charring, which makes it possible to calculate fire resistance by subtracting the charred area from the original cross-section together with a thin layer of heat-affected timber. Unprotected heavy timber structural components provide high fire resistance thanks to this charring characteristic, for example, much better than unprotected structural steel.

Encapsulation is the process of enclosing flammable materials with wood surfaces [23]. Although encapsulation can increase fire safety, not all fire issues in timber buildings are quickly resolved [24]. Full encapsulation requires that wood be encapsulated with enough layers of protective material to prevent any ignition or charring of wood in a complete burnout of the fire compartment, giving the same fire resistance as any non-combustible materials. Partial encapsulation slows down the development of fire on wood surfaces, but because there are fewer layers, they might fall off before fire burns out, leaving structural timber accessible to the advancing flames. In both cases, there is an aesthetic dilemma because building owners and users and their architects want to see wood, whereas fire engineers may need to cover it all from view.

Designing for burnout in a timber structure can only be done with absolute certainty by using full encapsulation, which prevents any structural timber from ever starting to char during the whole period of the fire growth, development, and decay. The necessary encapsulation relys on a number of variables:

- Intensity and length of the fire's burning period.
- Speed at which temperatures fall as a result of ventilation during the fire's decay phase.
- Ability of encapsulating materials.
- Intervention after fire has been put out.

An automated sprinkler system serves as the primary active fire defense strategy. The efficacy of the sprinkler system in high-rise timber buildings is the main concern for designers and code writers. The additional requirements for fire safety and fire resistance are low if sprinklers can be assured to function, but if for some reason they cannot, timber buildings fall into a distinct category and require extra care.

High-rise timber buildings must be designed with passive fire safety in mind on all levels, including egress, fire resistance, hidden cavities, etc. In order to protect the wood surfaces from fire, the focus is now on using encapsulation. Several choices include:

1. All exposed wood on walls and ceilings.
2. All walls are visible, but ceilings are covered.
3. All columns and beams are visible, and walls and ceilings are secure.
4. One layer of plasterboard covered in gypsum protects all wood (partial encapsulation).
5. Complete plaster board encapsulation, covering every piece of wood.

Designers of steel and reinforced concrete buildings work on the basis that if the compartmentation works and the structure is designed for adequate fire resistance, fire will eventually go out after the fuel is completely consumed, and the structure will cool to ambient temperatures after any heat remaining in the building has dissipated. This scenario is less certain for timber buildings because there will always be some fuel present in the wood components of the timber structure, leading to the possibility of slow charring which might continue long after the main fire has gone out. In this case, it is only possible to design a high-rise timber building with the same level of fire safety as a steel or reinforced concrete building if design for burn-out can be accomplished.

1.7 Building Height

Table 1.4 lists various building heights, ranging from low to high rise, along with recommendations for passive fire prevention based on sprinkler protection and building height. The definitions of the rows labeled "no sprinklers" and "normal sprinklers" in Table 1.4 are as normal [24]. The additional row for "special sprinklers" has been added to allow designers to take special precautions, such as a dedicated water tank in the building to ensure that the sprinklers have water, even if the

Table 1.4 Suggestions for passive fire protection depending on sprinkler protection and height of building

Height	Low-rise	Mid-rise	Tall	Very tall	High-rise
Stories	1–2	3–5	6–8	9–15	>15
Likely escape	Quick escape	Slow escape	Assisted escape	Assisted escape	Difficult escape
No sprinklers	Local areas exposed	No exposed wood	Not allowed	Not allowed	Not allowed
Normal sprinklers	Large areas exposed	Local areas exposed	No exposed wood	Full encapsulation	Full encapsulation
Special sprinklers	Large areas exposed	Large areas exposed	Local areas exposed	No exposed wood	Full encapsulation

street mains are destroyed by an earthquake or landslide, in which case a greater area of visible wood could be provided in high-rise timber buildings.

The options provided in Table 1.4 are only examples to illustrate the variety of alternatives. It will be vital for code writers in many nations to create standards that rationally represent these concepts as high-rise timber buildings gain in popularity. More investigations of the possibilities, including quantitative risk assessment, could aid in their further definition.

The efficacy of automated sprinkler systems in high-rise buildings is the main concern for designers and engineers. The additional fire safety standards are low if sprinklers are guaranteed to function, but if they are not, timber constructions fall into a unique category and require greater care than non-combustible building materials.

1.8 Conclusions

This chapter dealt with the theory and practice of sustainability assessment and the knowledge about high-rise timber buildings. A summary and the key findings of the chapter are presented in the following:

- Modern technology and engineered timber materials like glulam, CLT, and LVL enable the construction of high-rise timber buildings. Comparing modern timber constructions with steel and concrete ones reveals advantages. The only renewable building material that is now accessible is timber, which is also ecologically favorable.
- Timber is an excellent heat insulator, and its manufacture requires less basic energy (mostly from fossil fuels) than concrete and other construction materials. Because of the prefabrication of components, cheaper transport of goods, and shorter project length, building with timber is more economical.
- Multiple criteria assessment allowed coming to the conclusion that taller timber buildings are efficient from both economic and environmental perspectives as the highest buildings received higher ranking positions.
- Building designers who prioritize sustainability objectives have long recognized the value in timber's renewability and light carbon footprint. There has been discussion on environmental advantages or carbon savings from the new developing mass timber items utilized in buildings. The cost of investing in such low-to-zero carbon emission buildings is increasing.
- Based on commercial construction cost data from the RSMeans database, a mass timber building design is estimated to have 26% higher front-end costs than its concrete alternative. For mass timber buildings, there is not much historical construction cost data and little operational cost data; further research and data are thus required to generalize these findings.

Most of the researchers assessed the environmental impacts of the low-rise residential building. Very few researchers focused on high-rise commercial buildings.

Significant amounts of materials and energy are required for high-rise commercial buildings. Therefore, further research is needed in this field to achieve sustainable building construction.

This chapter has made a number of recommendations to address the present issues with fire resistance design for high-rise timber buildings. Depending on the height of the building and the dependability of the water supply for the required sprinkler system, different levels of protection or encapsulation of the wood structure should be offered.

References

1. Kuzmanovska, I., Gasparri, E., Monne, D. T., & Aitchison, M. (2018, August). Tall timber buildings: Emerging trends and typologies. In *2018 world conference on timber engineering*.
2. Börjesson, P., & Gustavsson, L. (2000). Greenhouse gas balances in building construction: Wood versus concrete from life-cycle and forest land-use perspectives. *Energy Policy, 28*(9), 575–588.
3. Oliver, C. D., Nassar, N. T., Lippke, B. R., & McCarter, J. B. (2014). Carbon, fossil fuel, and biodiversity mitigation with wood and forests. *Journal of Sustainable Forestry, 33*(3), 248–275.
4. Žegarac Leskovar, V., & Premrov, M. (2021). A review of architectural and structural design typologies of multi-storey timber buildings in Europe. *Forests, 12*(6), 757.
5. Lindholm, J., & Reiterer, S. M. (2017). Marketing innovative products in a conservative industry: A look at cross-laminated timber. Master's Thesis. https://www.diva-portal.org/smash/get/diva2:1120866/FULLTEXT01.pdf
6. Skullestad, J. L., Bohne, R. A., & Lohne, J. (2016). High-rise timber buildings as a climate change mitigation measure–A comparative LCA of structural system alternatives. *Energy Procedia, 96*, 112–123.
7. Quebec Wood Export Bureau (QWEB). (2015). *Wood construction technologies in Quebec, Canada*. Retrieved from http://www.quebecwoodexport.com/images/stories/pdf/Catalogue_AN_13-04-2016.pdf
8. Bergman, R., Puettmann, M., Taylor, A., & Skog, K. E. (2014). The carbon impacts of wood products. *Forest Products Journal, 64*(7/8), 220–231. https://doi.org/10.13073/FPJ-D-14-00047
9. Alliance for Buildings and Construction (GlobalABC). (2020). *Global status report for buildings and construction*. Available online: https://globalabc.org/news/launched-2020-global-status-report-buildings-and-construction. Accessed 18 Apr 2023.
10. The 17 Goals. Sustainable Development. https://sdgs.un.org/goals. Accessed 20 Apr 2023.
11. European Commission. *EU climate action and the European green deal*. Available online: https://climate.ec.europa.eu/eu-action/european-green-deal_en. Accessed 19 Apr 2023.
12. Design Build Network. (2019). *The world's tallest wooden buildings. The world's tallest wooden buildings*. worldconstructionnetwork.com
13. Tupėnaitė, L., Žilėnaitė, V., Kanapeckienė, L., Sajjadian, S. M., Gečys, T., Sakalauskienė, L., & Naimavičienė, J. (2019). Multiple criteria assessment of high-rise timber buildings. *Engineering Structures and Technologies, 11*(3), 87–94.
14. WSP. (2018). *Tall timber: How high can CLT go?* Retrieved from https://www.wsp.com/en-RO/insights/tall-timber-how-high-can-clt-go
15. Balasbaneh, A. T., & Bin Marsono, A. K. (2017). Proposing of new building scheme and composite towards global warming mitigation for Malaysia. *International Journal of Sustainable Engineering, 10*(3), 176–184.

16. Rios, F. C., Grau, D., & Chong, W. K. (2019). Reusing exterior wall framing systems: A cradle-to-cradle comparative life cycle assessment. *Waste Management, 94*, 120–135.
17. Al-nassar, F., Ruparathna, R., Chhipi-Shrestha, G., Haider, H., Hewage, K., & Sadiq, R. (2016). Sustainability assessment framework for low rise commercial buildings: Life cycle impact index-based approach. *Clean Technologies and Environmental Policy, 18*, 2579–2590.
18. Malabi Eberhardt, L. C., van Stijn, A., Nygaard Rasmussen, F., Birkved, M., & Birgisdottir, H. (2020). Development of a life cycle assessment allocation approach for circular economy in the built environment. *Sustainability, 12*(22), 9579.
19. Caldas, L. R., Saraiva, A. B., Lucena, A. F., Da Gloria, M. H. Y., Santos, A. S., & Toledo Filho, R. D. (2021). Building materials in a circular economy: The case of wood waste as CO2-sink in bio concrete. *Resources, Conservation and Recycling, 166*, 105346.
20. Tupenaite, L., Zilenaite, V., Kanapeckiene, L., Gecys, T., & Geipele, I. (2021). Sustainability assessment of modern high-rise timber buildings. *Sustainability, 13*(16), 8719.
21. Framework. (2020). *Framework.* https://www.frameworkportland.com/. Accessed 19 Apr 2023.
22. Gu, H., Liang, S., & Bergman, R. (2020). Comparison of building construction and life-cycle cost for a high-rise mass timber building with its concrete alternative. *Forest Products Journal, 70*(4), 482–492.
23. Buchanan, A. H., & Abu, A. K. (2017). *Structural design for fire safety*. Wiley.
24. Buchanan, A. H. (2017). Fire resistance of multistorey timber buildings. In *Fire science and technology 2015: The proceedings of 10th Asia-Oceania symposium on fire science and technology* (pp. 9–16). Springer Singapore.

Chapter 2
Ultralight Membrane Structures Toward a Sustainable Environment

Alessandro Comitti, Harikrishnan Vijayakumaran, Mohammad Hosein Nejabatmeimandi, Luis Seixas, Adrian Cabello, Diego Misseroni, Massimo Penasa, Christoph Paech, Miguel Bessa, Adam C. Bown, Francesco Dal Corso, and Federico Bosi

2.1 Introduction

The climate change experienced by our planet has alarmingly escalated in recent decades due to anthropogenic contributors to environmental degradation. The present scale of this global emergency demands urgent and immediate remedial measures to ensure that a safe biosphere prevails for future generations of humanity and

A. Comitti · L. Seixas
Department of Mechanical Engineering, University College London, London, UK

CAEMate s.r.l., Bolzano, Italy
e-mail: a.comitti@ucl.ac.uk; luis.seixas.20@ucl.ac.uk

H. Vijayakumaran
Department of Material Science and Engineering, Delft University of Technology, Delft, Netherlands

Tensys Ltd., Bath, UK
e-mail: h.vijayakumaran@tudelft.nl

M. H. Nejabatmeimandi
Tensys Ltd., Bath, UK

Department of Civil, Environmental and Mechanical Engineering, University of Trento, Trento, Italy
e-mail: mh.nejabatmeimandi@unitn.it

A. Cabello · A. C. Bown
Tensys Ltd., Bath, UK
e-mail: adrian.cabello@tensys.com; adam.bown@tensys.com

D. Misseroni · F. Dal Corso (✉)
Department of Civil, Environmental and Mechanical Engineering, University of Trento, Trento, Italy
e-mail: diego.misseroni@unitn.it; francesco.dalcorso@unitn.it

M. Penasa
CAEMate s.r.l., Bolzano, Italy
e-mail: massimo@caemate.com

A. Bahrami (ed.), *Sustainable Structures and Buildings*,
https://doi.org/10.1007/978-3-031-46688-5_2

nature. As a result, the United Nations (UN) has identified climate change as a central challenge faced by the planet, as it intrinsically affects all the *Sustainable Development Goals* (SDGs) set forth by the UN. Among the 17 SGDs stated by the UN, building construction practices have an impactful role in achieving SDG 11 – *sustainable cities and communities*, and SDG 12 – *responsible consumption and production* [1], which highlights the need for a paradigm shift in the construction industry to achieve the clean energy transition.

Building constructions and operations show the highest environmental footprint, with 36% of global energy consumption and 39% of carbon dioxide (CO_2) emissions, the latter greater than transportation (33%) and industrial activities (29%) [2]. A recent report by the United Nations Environment Programme [3] demonstrates that the carbon footprint of constructions is increasing, with 28% of buildings-related CO_2 emissions finding their roots in the use of materials and that the demand for buildings and floor area is growing and expected to double by 2060. Consequently, the requirement for materials will remarkably increase in urban contexts, primarily in Asia and Africa. Under these circumstances, innovative building technologies employing low-carbon materials are of paramount importance in embodied carbon reduction to lower construction-related CO_2 emissions through (i) resource-efficient lightweight building designs, (ii) waste reduction via reuse and recycling, (iii) lifetime extension, and (iv) minimal transportation. Hence, the main challenge in the building sector is seeking and implementing novel construction technologies.

A feasible solution toward the achievement of a sustainable built environment is offered by *membrane*, or *tensile*, structures. This construction type aims to bear the external loads through structural elements acting under tension, differently from the load-bearing mechanisms displayed by traditional structures, namely compressive states for arches, bending-dominated states for frames, and compressive/tensile states for trusses. Consequently, the heavyweight and stiff constructions realized in the past through a considerable amount of concrete, steel, stone, and timber materials could, in some cases, be replaced by lightweight and flexible tensioned membranes. Throughout the centuries, the ratio between the self-weight of a permanent structure and the load it carries, defined as γ, has been decreasing, reaching approximately unity with the advent of structural steel, and falling below 1.0 in the case of tension structures, as depicted in Fig. 2.1. A lower ratio implies a decrease in structural weight and more effective use of building materials. Hence, leveraging on an efficient load-bearing mechanism, membrane structures require a reduced amount

C. Paech
sbp – schlaich bergermann partner, Stuttgart, Germany
e-mail: c.paech@sbp.de

M. Bessa
School of Engineering, Brown University, Providence, RI, USA
e-mail: miguel_bessa@brown.edu

F. Bosi (✉)
Department of Mechanical Engineering, University College London, London, UK
e-mail: f.bosi@ucl.ac.uk

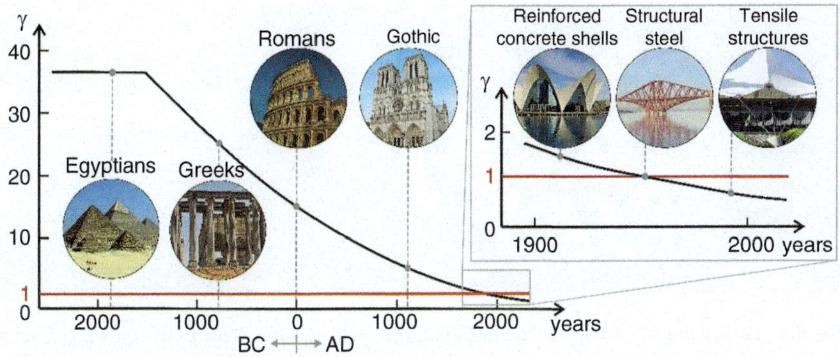

Fig. 2.1 Evolution throughout centuries of ratio γ between weight of a structure and carried load, indicating efficiency of membrane structures [4, 5]

of materials, which in turn reduces the energy and emissions related to their production and transportation.

The efficiency of tensile structures was already exploited by many ancient civilizations around the world. The first examples are the masts of Egyptian sailboats, the movable roofs of Roman amphitheaters, and the tents of Bedouins and Navajo tribes. The possibility to build large-scale tensile/membrane structures has been reached only in the modern era as a consequence of scientific and technological advances. The first large-scale tensile buildings were developed in the second half of the twentieth century, especially for exposition structures. Among the many realizations, the Olympiastadion (Munich Olympic Stadium) by the architect Frei Otto is the most iconic tensile structure of that era [6]. However, at that time, these solutions were limited to temporary installations because of the undeveloped technologies in the material field. Therefore, tensile structures were not yet attractive in terms of durability and sustainability. Nowadays, the technology advances in lightweight materials allow for designing tensile structures with a 30-year lifespan, thanks to improved coatings or foils with superior environmental weathering resistance [7].

As a result of this improved durability, membrane structures are currently employed in a broad spectrum of building applications, such as claddings, roofing, and facades for fairs, exhibitions, and stadia, realizing beautiful envelope designs while pursuing the optimization of resources, Fig. 2.2.

This chapter aims to provide the basics of lightweight membrane structures and the evidence of their role toward green and sustainable constructions. The mechanical principles defining the efficiency of the tensile load-bearing mechanisms in relation to weight and material savings are presented in Sect. 2.2. The technological aspects inherent to the realization of lightweight membranes and the importance of accurate mechanical modeling for optimal material use and safe design are discussed in Sect. 2.3. A quantitative assessment of sustainability aspects is addressed in Sect. 2.4. Each of these sections contains a description of current challenges and

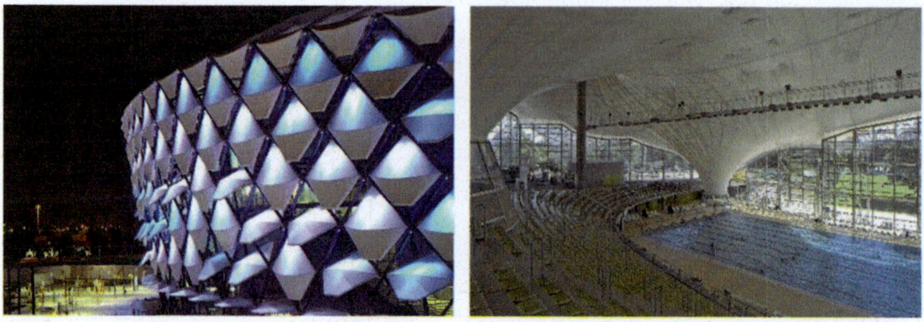

Fig. 2.2 Two examples of tensile structures in large-scale constructions. Left: Glass/PTFE membrane panels for external shading at Hazza Bin Zayed Stadium (© Christoph Paech/schlaich bergermann partner). Right: Membrane roof of Olympic Aquatic Centre, Munich (© Michael Zimmernann/schlaich bergermann partner)

opportunities in the relevant engineering aspect. Lastly, conclusions on lightweight membrane structures are drawn in Sect. 2.5.

2.2 Engineering Design of Ultralightweight Membrane Structures

Membrane structures are an ensemble of lightweight structural elements that combine the principles of aesthetic architecture, material optimization, and structural efficiency. They are advantageous in scenarios where the design has to accommodate large unsupported spans with minimal weight. By building better with less material, environmental benefits in the form of reduced energy usage and carbon emissions during production, transportation, and installation could be accrued, while simultaneously providing a cost-effective structural engineering solution.

2.2.1 Structural and Material Efficiency Through Tensile State

Membrane structures, analogous to their parent class of tensile structures, are designed with the principle of maximum structural efficiency at their core. Bending and torsion are disadvantageous load-bearing mechanisms as the material near the neutral plane is mostly unused. On the contrary, tensile forces generate a constant stress distribution normal to the cross-section, efficiently using all material throughout the thickness and resulting in a significant weight reduction [8]. By remaining in the state of pure tension, membrane structures not only achieve low weight-to-load ratios but also encounter a reduction in instabilities and stress localizations typical of compression [9, 10] and bending-dominated structures [11, 12].

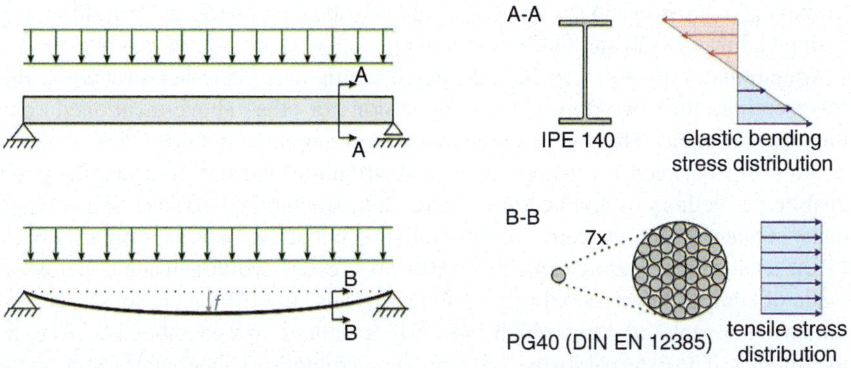

Fig. 2.3 Comparison between bending (top) and tension (bottom) load-bearing elastic mechanisms for a pin-supported structure subjected to the same uniformly distributed load

Table 2.1 Engineering satisfiability and weight comparison of pinned I-beam and cable solutions, associated with bending and tensile load-bearing mechanisms, respectively

Description	I-beam	Cable
Span	4.0 m	
Uniform vertical load	10.0 kN/m	
Ultimate limit state utilization factor	0.65	
Material	Hot rolled steel	Galfan coated steel
Design element	I-section IPE 140 mm (S355)	Spiral strand 1×37, Ø 20.1 mm (PG40 [13])
Weight per meter	12.9 kg/m	1.9 kg/m

To elucidate the efficiency of tension over bending, consider an l = 4.0 m pin-supported span subjected to a uniformly distributed load q = 10.0 kN/m, as illustrated in Fig. 2.3. An efficient ultimate limit state (ULS) design in bending can be obtained by using a steel (S355) I-beam of 140 mm height (elastic section modulus S = 77.32 cm^3, sectional area A = 16.4 cm^2), which results in a weight-per-length ratio of 12.9 kg/m. In contrast, the same load can be borne under pure tension by a 20.1 mm diameter steel cable of open spiral strand cross-section (Galfan-PG40 [13], limit design force $F_{u,d} = 222$ kN), assuming the same utilization factor α at the ultimate state and a maximum deflection f equal to the height of the I-beam, such that the tensile force in the cable is $T \simeq ql^2/(8f) = \alpha F_{u,d}$. Such tensile solution has a weight-per-length ratio of 1.9 kg/m, resulting in 85% weight savings. The comparison is detailed in Table 2.1, which demonstrates through the weight-per-length ratio, how a tensile solution with a flexible cable better utilizes the material compared to the rigid I beam. Figure 2.3 compares the cross-sectional stress distribution in the two solutions.

It should be noted that additional mass savings can be achieved if the maximum deflection f is increased, thus decreasing the maximum tensile force in the cable. For example, $f = 200$ mm and $f = 400$ mm would provide 93% and 97.5% weight savings (using Galfan's PG20 and PG5), respectively.

Membrane structures can be considered continuous cable net structures, thus representing the two-dimensional extension of the above-mentioned one-dimensional cable. They are designed to exploit engineering knowledge about the relationship between the nature of load distribution and the deformation states, enabling a plethora of architectural shapes and structural approaches. The components of a membrane structure are assembled such that the loads are primarily borne by the tensile load-bearing elements: (i) the prestressed two-dimensional membrane made of composite/woven fabric or polymeric foils, and (ii) the pre-tensioned one-dimensional cables and ties which form ridges, valleys, and edge boundaries to the membranes. The external loads acting on the membrane element manifest as membrane stresses, as displayed in Fig. 2.4, while those acting on the cable and tie elements take the form of axial forces. These tensile load-bearing elements then transfer the loads to the structural support framework composed of trusses, masts, and beams, which are designed to withstand compression, bending, shear or torsion loads.

The efficient load transfer mechanism of membrane structure relies on prestressing the tensile load-bearing elements, which are inherently flexible. If these elements are not pre-tensioned during installation, they would go slack and undergo considerable displacements, becoming highly susceptible to structural instabilities such as flutter and wrinkling [14, 15]. Based on the method of pre-tensioning of the membrane elements, we can distinguish them as *boundary-tensioned* and *pneumatic* membrane structures. The load-bearing mechanisms of the two categories are different, as are the engineering principles guiding their design. A resource-efficient

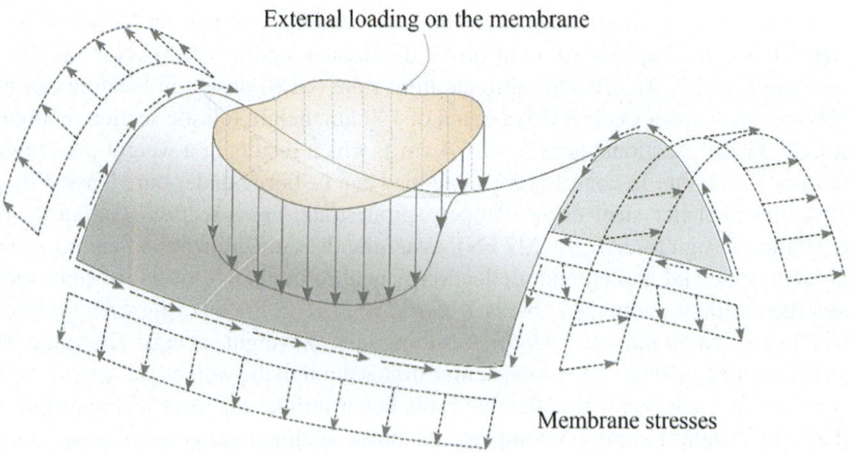

Fig. 2.4 Membrane systems transfer arbitrary loads applying on them as membrane stresses that act along tangent plane to mid-surface of membrane

design of these structures critically relies on understanding the nuances in the implications of these differences.

2.2.2 Boundary-Tensioned Membrane Structures

Boundary-tensioned membrane structures (Fig. 2.5) are prestressed by stretching the membrane elements along its boundaries, which are made of either flexible tension cables or rigid frames/beams.

A fundamental difference between membrane structures and typical civil engineering structures made with concrete, steel, or timber is that the load-carrying capacity of the former arises from curvature or form adaptation. To appreciate how curvature or form adaptation works, it is helpful to understand the concept of constrained minimal surfaces and the weighted catenary mechanism.

Consider a cable or a rope hanging from two support points (Fig. 2.3). The idealized shape attained by a hanging cable or rope under its self-weight falls under the general class of *weighted catenary* [17]. Untensioned cables and ropes form such a curved shape because, unlike beams, their cross-section has negligible bending stiffness. The U-like form adaptation transforms the self-weight into tension in the cable or rope and transfers it to the support as a normal reaction force. It is also worth noting that, due to energy principles, the form of the catenary at equilibrium corresponds to that of least potential energy. This principle applies also in the presence of pre-tensioned state or hanging weights. The deformed state adapts to achieve the equilibrium shape, while the external loadings act as added tension along the structure.

Curvature or form adaptation in membranes works in a similar way, except that the form adaptation is sought by a surface. The curved surfaces formed by membranes under external loads, including the simplest load case of self-weight, fall into the class of constrained *stable minimal surfaces* [18]. It implies that although the

Fig. 2.5 Example of a boundary-tensioned structure: Millenium Dome (O2 Arena), London (left) [16]; McArthurGlen Designer Outlet Village; architecture: Richard Rogers; engineering and fabrication: Buro Happold, Tensys Ltd., Architen Landrell (right)

designers can define the perimeter and support points for the membrane structures as constraints, the form of the structures is bounded by an envelope of limited possibilities, which depend on the design prestress. Thus, the form of a membrane structure has to be found through an iterative approach considering large deformations. This process differs from the technique used for typical concrete, steel, or timber structures, where the designer performs structural analyses under small deformation assumptions.

The process of deriving the form of the tensile elements of membrane structures is known as form-finding. A demonstrative example of form-finding in a membrane art installation is depicted in Fig. 2.6.

Since the structural support system is designed in congruence with the tensile elements of the membrane structure, form-finding plays a pivotal role in the overall structural optimization and weight savings. The geometrically nonlinear response of tensile structures demands reliance on computational simulations and specific

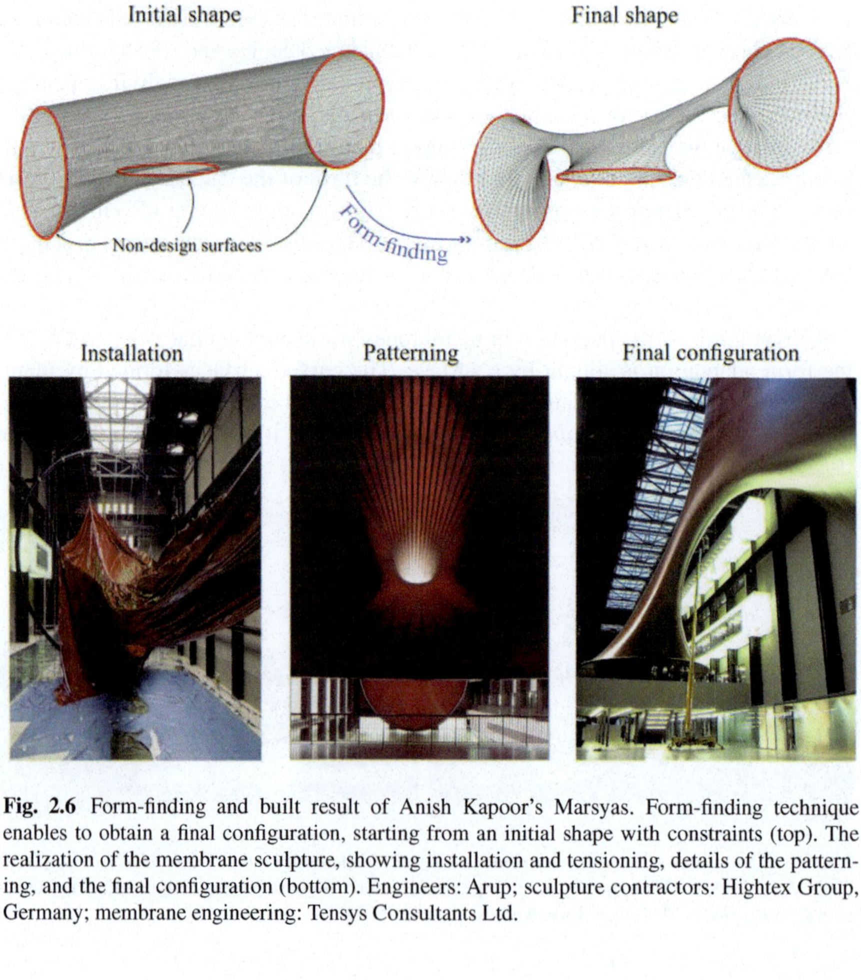

Fig. 2.6 Form-finding and built result of Anish Kapoor's Marsyas. Form-finding technique enables to obtain a final configuration, starting from an initial shape with constraints (top). The realization of the membrane sculpture, showing installation and tensioning, details of the patterning, and the final configuration (bottom). Engineers: Arup; sculpture contractors: Hightex Group, Germany; membrane engineering: Tensys Consultants Ltd.

numerical methods such as transient stiffness method [19], force density method [20], or dynamic relaxation method [21] for form-finding. Physical model-based approaches involving soap-film, fabric, or paper [9, 22] for the derivation of membranes form are also prevalently used for benchmarking the computationally evaluated forms.

An additional process, called patterning, is needed to design and install these structures. Although patterning is uncommon in other construction techniques, it is a necessary step in membrane structural design to determine how the shape obtained through the form-finding process will be realized. In order to perform this step, the membrane design shape must be traced back to an unstressed configuration and then decomposed into flat parts, approximating the 3D shape into bidimensional elements. During installation, the pieces of cut membranes from the raw sheets are joined mechanically or through welding. Once the complete membrane has been tailored, its installation requires careful procedures by skilled operators, so that prestress is gradually applied, and the product handling complies with the material specifications.

2.2.3 *Pneumatic Membrane Structures*

Pneumatic membrane structures are predominantly used as pressurized cushion-type cladding elements, air beams and façades, as demonstrated in Fig. 2.7. They are also popularly employed for designing inflatable event tents, temporary structures, art and architectural installations, as they offer effective solutions at minimal material weight and cost.

As opposed to boundary-tensioned membrane, pneumatic structures are composed of doubly or multi-layered panels prestressed by an internal inflation pressure. As a result, this structural configuration provides additional stiffness against bending-type loads due to their pressurized nature.

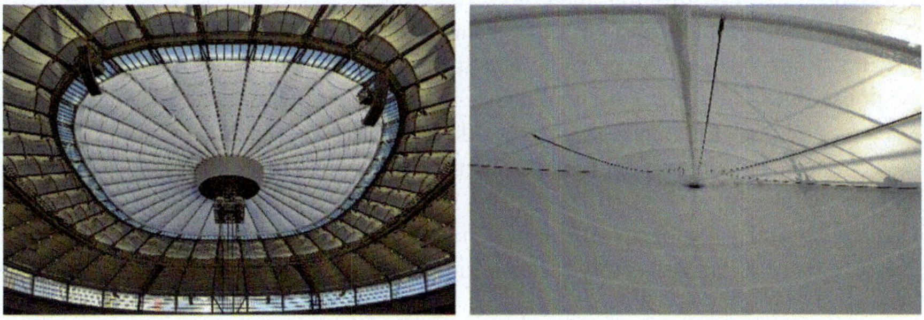

Fig. 2.7 Example of a pneumatic structure with cushion realized in PTFE-coated PTFE weave (left) and internal view of inflated cushion (right). BC Place Stadium, Vancouver, CA (© Christoph Paech/schlaich bergermann partner)

Furthermore, the form-finding process differs in inflated pneumatic membrane structures because the thin sheet's patterning and seams principally dictate their form. A form-finding exercise is usually performed to verify whether the inflated structure's final shape under the prescribed internal pressure is in accordance with the desired form. An adjustment in the design dimensions of the pressurized membrane panels follows by iterating this step until the final inflated shape is satisfactory. As described in boundary-tensioned membranes, the patterning step and complex installation procedure are also performed for pneumatic membranes, where each inflated cushion is patterned individually.

2.2.4 Structural Design Optimization Challenges and Opportunities

Although tension structures are based on an efficient load-carrying mechanism, design improvements could be achieved through an advanced understanding of the behavior of design materials and pertinent optimization techniques during structural design. Such comprehensive approaches could be a powerful tool at the initial, conceptual, and design stages of the membrane structure and can potentially identify efficient design states that are even beyond a design engineer's imagination [23].

Optimization strategies targeting structural geometry, component dimensions, shape, and topology are topics of active research in aerospace, mechanical, and civil engineering disciplines [24, 25]. In recent years, the accelerated advancements in the field of machine learning have influenced novel strategies for structural optimization, such as neural reparameterization [26], neural density representation [27], and Bayesian structural optimization [28]. However, literature on structural optimization methods applied to membrane structures are scarce, indicating the presence of a knowledge gap. Drawing parallels from the aforementioned disciplines, a pursuit to fill this knowledge gap by employing classical and machine learning-driven optimization methods could result in substantial economic and environmental gains by means of further efficient material utilization.

While these structural optimization methods primarily assume elastic response, the materials used in membrane structures do not retain such simple material response throughout their life cycle. The membrane materials undergo phenomena such as yielding, time and temperature-dependent responses, fatigue, and damage, which can severely impact the mechanical performance of the material and that of the overall structure [29, 30]. Therefore, mathematical models that accurately predict the complex responses of the materials used in membrane constructions must be developed as an accessory to computer-aided design and optimization of the overall structure.

2.3 Membrane Materials

To achieve sustainable design solutions through membrane technology and fully exploit its advantages, the structural efficiency described in the previous section must be paired with a detailed understanding of the intrinsic properties and behaviors of the materials used. Membrane materials need to be flexible to achieve an adequate displacement and a pure tensile stress state that complies with the forms imposed by the boundary conditions. Hence, the bending stiffness should be negligible. This is obtained using relatively soft composites or homogeneous materials, whose stiffness is more than one order of magnitude lower than concrete, and of a reduced thickness, generally lower than 1 mm, consequently producing a lightweight solution for the envelope. Membrane structures are currently realized by employing two leading technologies: *fabrics* and *foils*, where the latter is mainly produced with polymers.

2.3.1 Fabrics and Foils

The load-carrying base cloth of woven fabrics is formed from yarns combined in different ways to create a matrix of interlaced threads. The threads are engaged in two orthogonal directions, called warp and weft, through various techniques (two examples of these are displayed in Fig. 2.8). The weaving techniques used to produce the base cloth cause a waviness in the trajectories of the fibers, different in the two planar directions, resulting in technique-dependent material stiffness and nonlinear orthotropic behavior. In most fabrics, the warp direction is stronger and stiffer than the weft one [7].

Textile fabrics are usually coated to protect the base cloth from environmental weathering as well as to ensure the sealing against water and air. The coating layer

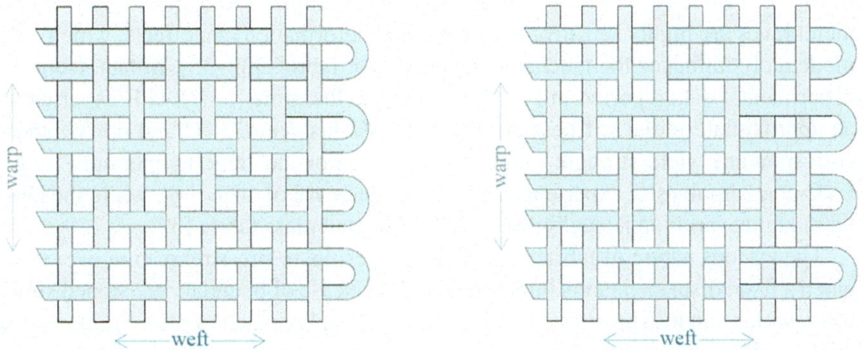

Fig. 2.8 Two most common arrangements of threads in a fabric: plane wave (left) and basket wave (right)

also plays a key role as a structural component, as it enhances the in-plane shear stiffness of the fabric and transfers the stresses between different panels through welded joints. The two most widely used families of woven fabrics are PVC-coated polyester and PTFE-coated glass fibers. Their main differences are the durability, which is around 20–25 years for the former and 30 years for the latter, and the behavior after yield. In fact, PVC-coated polyesters are very flexible and comply with any shape a modern envelope might request. On the contrary, PTFE-coated glass fiber fabric is rather brittle because of the glass material, thus requiring additional care during the installation phase [7].

The category of foils mainly refers to the material ethylene–tetrafluoroethylene (ETFE), a semicrystalline thermoplastic polymer that is extruded to produce a homogeneous foil. The expected lifespan of an ETFE solution is 30 years. ETFE is very similar in its chemical structure to PTFE, indicating high UV resistance, ductility, self-extinguishing properties and a light transmission superior to that of glass. The last property is especially fascinating for architectural applications since the material provides natural light to the interiors of the building, offering a lighter and greener alternative to glass. This results in large weight savings in the envelope and supporting structures, thus reducing the environmental footprint and the energy required for production (~10 times lower than glass) and installation (24–70% less than glass). Lately, researchers are also exploring the possibility of embedding photovoltaic cells during production, with promising results for solar energy collection and the clean energy transition [30].

2.3.2 Thermomechanical Response of Structural Membranes

Membrane materials are highly nonlinear and undergo large deformation. Moreover, the polymeric nature causes their response to be dependent on *temperature* and *time*. High strain rates and low temperatures increase the stiffness and strength, while low strain rates and high temperatures soften the response [31], as illustrated in Fig. 2.9 for ETFE foils as a representative example. More specifically, ETFE membranes are highly sensitive to thermal and deformation rate effects, which only moderately influence the mechanical response of fabrics [32–35]. In fact, from the experimental data reported in Fig. 2.9 and in the literature [34], ETFE experiences an approximate initial stiffness decrease of 35% between 23 °C and 60 °C, and a corresponding yield strength reduction of 40%. For the same temperature range, PVC-coated polyester experiences a stiffness decrease of 20% [36], while the elastic modulus variation is negligible for PTFE-coated glass fabric [37].

The viscous nature of fabrics and foils also results in stress relaxation over time, when a fixed displacement is imposed. This is the case of boundary-tensioned structures, which can lose prestress over time due to this phenomenon [39]. Similarly, the dual condition of constant applied stress realizes a continuous increase of strain over time. Such creep conditions can occur in pneumatic membrane structures,

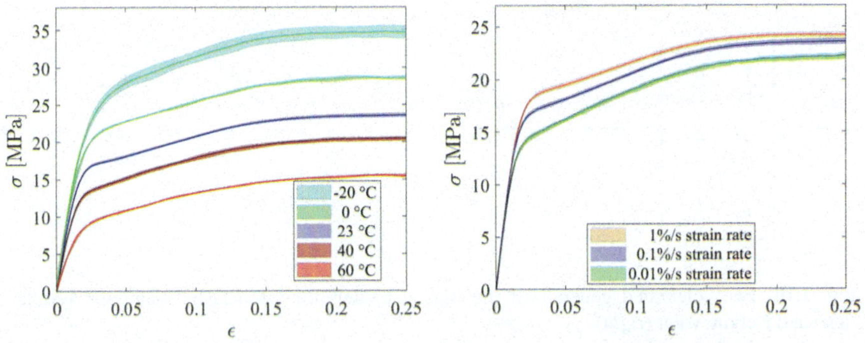

Fig. 2.9 Engineering uniaxial stress-strain curves of ETFE foils, experimentally measured within LIGHTEN project [38]. Response for (left) different temperatures at a constant strain rate of 0.1%/s and (right) for different strain rates at a constant temperature of 23 °C

where the constant inflation pressure generates a stress state dependent on the shape of the cushion. If the material continues to strain due to its viscosity, the shape will change, causing the prestress, and hence the stiffness, to decrease.

2.3.3 Constitutive Modeling Challenges and Opportunities

All the above-mentioned temperature and time effects strongly affect most membrane material responses and must be accounted for, in order to achieve an optimal design. However, these features are rarely considered by designers at present, due to the lack of construction codes for membrane structures [40] and the suggestion to adopt linear elastic models in pre-standard documents [41]. The use of such simplified approaches can result in overdesign, employing unnecessary material, or unsafe design, as reported by Cabello and Bown [42]. The lack of confidence in the available design tools and material models is hindering the use of membrane technology, weakening the impact that tensioned structures could have in reaching sustainable construction practices.

To fully exploit the recognized potential of membrane structures in the environmental cause and the versatility of their applications, comprehensive thermo-visco-elasto-plastic constitutive models must be developed. In particular, the definition of the time and temperature dependence of the mechanical properties and yielding [43] of membrane materials is of great interest among the engineering community [34, 42]. These features also influence the representation of design load cases, whose expressions should include temperature condition and loading velocity. Comprehensive experimental campaigns at multiple conditions represent the stepping stone for the development of accurate material constitutive relations through different modeling strategies, including data-driven approaches, which are part of current research efforts [38].

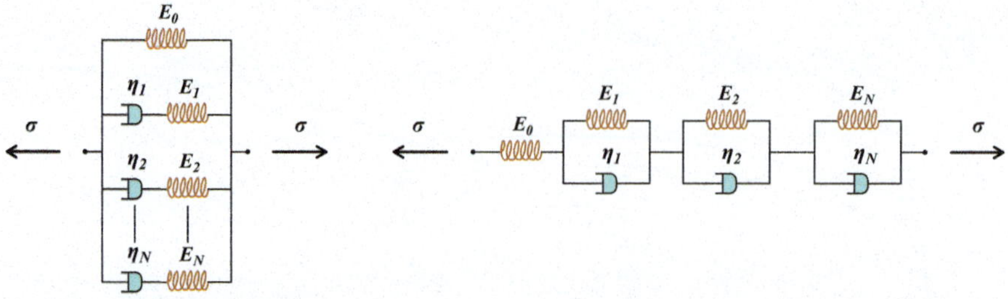

Fig. 2.10 Two celebrated generalized rheological models for viscoelastic materials: Wiechert (left) and Kelvin-Voigt (right)

Examples of constitutive relations capable of overcoming the current linear elastic approaches in order to capture the complex nonlinear and time-dependent membrane viscoelastic response are given by rheological models. They are defined through a combination, in series or parallel, of N linear dashpots of viscosity η_i (i = 1, .., N) and N + 1 springs of stiffness E_i (i = 0, ..., N), as displayed in Fig. 2.10 [44].

For a generalized Kelvin-Voigt model, the time-dependent relationship between strain $\varepsilon(t)$ and stress $\sigma(t)$ can be expressed by defining $D_i = E_i^{-1}$ and $\tau_i = \eta_i/E_i$ and by applying the Boltzmann superposition principle as:

$$\varepsilon(t') = \int_0^{t'} D_0 + \sum_{i=1}^{N} D_i \left(1 - e^{-\frac{t'-s}{\tau_i}} \right) \frac{d\sigma(s)}{ds} ds$$

where the term $D_0 = E_0^{-1}$ represents the instantaneous stiffness of the material. Since temperature affects the viscosity of materials, this dependence can be incorporated into the model of thermo-rheologically simple materials through the time-temperature superposition principle. Recent contributions in the literature report some attempts to model membrane materials with rheological models [45, 46]. However, the results for building construction materials are still not sufficiently developed to allow their use in the design of membrane structures [47, 48].

2.4 Sustainability of Membrane Structures

Although common sense would relate polymers to materials with high environmental impact and production cost, these two aspects are less crucial for polymeric membrane structures when compared to traditional constructions. Sustainability of membrane structures is addressed in the following in terms of embodied energy, material consumption, and recyclability.

Table 2.2 Energy required to manufacture the same volume of different materials, normalized by corresponding value for steel [49]

Material	Relative energy per unit volume	Material	Relative energy per unit volume
Steel	1	Polystyrene	0.14
Aluminum	0.68	HDPE	0.10
Nylon	0.23	PVC	0.10
Polycarbonate	0.20	LDPE	0.08
Acrylic	0.19	Polypropylene	0.08

2.4.1 Embodied Energy and Material Consumption

The relative energy needed to produce the same volume of different materials is reported in Table 2.2, using the energy consumption data to produce sheets of different materials. The table shows that the manufacturing cost for polymers is less than 25% of that for steel [49].

In particular, the comparison of embodied energy in membranes is more properly referred to a square meter of envelope elements. For example, ETFE elements have an embodied energy that varies between 27 and 210 MJ/m^2 (overestimating the weight to 1 kg/m^2). This value depends on the solution chosen among a single foil or a multi-layered cushion, and the factors considered in the energy calculations [50, 51]. Comparing ETFE to the most common transparent cladding technology in roofs and facades, float glass, a 6 mm thick panel has an embodied energy of 300 MJ/m^2 [50]. The benefit in terms of embodied energy of ETFE solutions would be even higher if the overall structural system is considered, as ETFE enables lighter supporting structures. In this regard, three case studies of transparent roofing construction collected from the literature [52, 53] are displayed in the histogram of Fig. 2.11. For each of them, the weight per unit area is plotted under the hypothesis of using either glass or EFTE as cladding materials. The efficiency of the tensile structure system is evident as it leads to an overall weight saving in the construction, with a reduction of the material consumption varying from 45% to 85% with respect to the corresponding glass roofing installation.

Assessment of further sustainability aspects for ETFE and fabric structures is also available in the literature. In particular, stadium facades and atria roofing designs with tensile structures have been compared to traditional technologies in terms of carbon footprint [54], primary energy consumption [52], or completing a life cycle impact assessment [53]. The consensual conclusion is that structures with membrane elements are more environmental-friendly than traditional constructions.

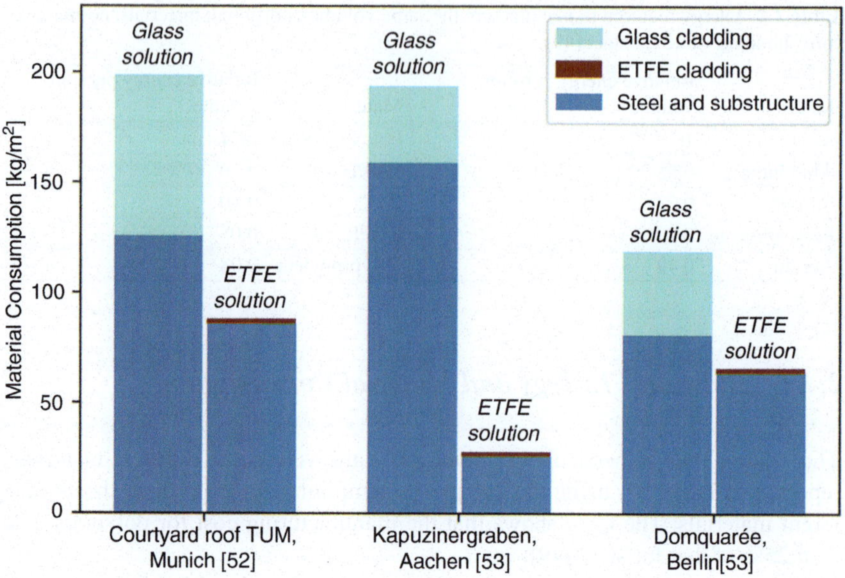

Fig. 2.11 Material consumption comparison between glass and ETFE solutions for transparent roofing for three different case studies [52, 53]

2.4.2 *Recyclability*

In addition to the aspects of low embodied energy and material consumption, the sustainability of tensile structures is further provided by the reuse and recycling of their components, membranes, and substructures. Indeed, an emerging trend is to reuse fabrics that have reached their lifespan, finding a second life in a less demanding environment. Although reusing can represent a meaningful way to reduce the carbon footprint of the membranes, recyclability currently offers more alternatives. In fact, it is well known that thermoplastics lack strong bonds between the polymeric chains, so they can be recycled. ETFE [30] and PVC-coated polyester membranes [55] are eligible for this process and thereby reduce the amount of their embodied energy because some of the production steps necessary for the virgin material can be circumvented during the process of recycling.

Traditional fabric materials are difficult to recycle because the textile composites need to be separated in their components, woven and coating: that is the reason why a textile membrane like PTFE-coated glass is not recycled. Nevertheless, the manufacturers of foils and fabrics are considering recycling options and are committing to more sustainable production. Examples are the Texyloop [55] project of Serge Ferrari on PVC-coated polyester membranes, which aimed to recycle cut-off and unused fabrics to produce new raw materials, and a recent project run by the startup Polyloop [56] that aims to recycle PVC from PVC-composite materials.

Because of its homogeneous nature, the recycling process of ETFE is relatively simple, yet essential in reducing the production of dangerous substances for ozone layer depletion, such as the R11 (trichlorofluoromethane) and R22 (difluorochloromethane) emission during its polymerization [53]. At the current state of the art, the Environmental Product Declarations of the ETFE cushion system of Vector Foiltec, a market leader in ETFE systems design and building [57, 58], reports that the material is recycled, either to produce new foils or to realize other ETFE components such as pipes and valves, mainly used for pneumatic cushions [59]. The recycling of ETFE foil as a new material is indicated to reduce the R11 emission by 47%, while it reduces global energy consumption[1] by 14% [57, 58]. ETFE foil also outperforms glass in the recyclability aspect, since the glass used for buildings is almost non-recyclable because of the difficulties in removing coating layers, and it demonstrates shallow (5%) energy savings [60].

2.4.3 Thermal Properties Challenges and Opportunities

The current major drawback of membrane technology, related to sustainability, is its thermal performance. Although the lightness of the cladding system enables most of the advantages of the solution, the reduced thickness of the membrane materials causes its thermal conductivity to be higher than the traditional building technologies [61]. For example, an ETFE cushion has a thermal transmittance in a range between 2.9 and 1.4 W/m^2K, depending on the number of layers [62], while modern glazing systems reach values that range from 2 [61] to 1.1 W/m^2K, depending on the surface treatments [63]. Nowadays, ETFE's low thermal performance limits its application in tensile structures mainly to atria, sports halls, stadia roofing, squares, industrial buildings, and other constructions where thermal requirements are not priorities. Several strategies can be adopted to improve and mitigate this drawback, such as the use of multi-layered membrane structures or cushions. These solutions provide additional separation layers from the outer environment and take advantage of the insulation given by the air layer [64]. Moreover, the use of airflow, either natural [65] or generated, toward the surface of the membranes, has been illustrated to improve the energetic performance [66] and has been successfully employed in large-scale projects such as the Khan Shatyr in Astana [67]. Additional improvements can be obtained by using different colors, reflective coatings, and printed patterns, especially on ETFE transparent solutions, to tailor the thermal radiation performances to the specific project demand [30].

[1] Considering the sum of the PENRT (total use of non-renewable primary energy resources) and PERT (total use of renewable primary energy resources) indexes of the life cycle assessment [57, 58].

2.5 Conclusions

Membrane structures display great sustainability potential due to an efficient load-bearing mechanism realized through a reduced amount of highly recyclable materials with low embodied energy. The maturity reached in engineering design and material aspects defines this construction type as a promising sustainable solution for a broad range of applications in building engineering, encompassing large-scale facilities, and correspondingly as the opportunity to reduce the impact of the construction industry on our planet and contribute to the goals set by the COP21 Paris Agreement.

Tensile membrane structures have been shown to be a reliable construction technology, with satisfying durability and versatility in adapting to multiple envelope requirements, from small to large scale, across different climates. The latest advancements discussed and the recycling possibilities are expected to anticipate a wider spread of the technology to obtain lighter buildings and more resource-conscious designs in the near future. The main opportunities and challenges to be faced for pursuing the next step forward in membrane structures sustainability have been described at the end of each previous section and include structural design optimization, thermo-visco-elasto-plastic constitutive modeling, material recycling, and thermal properties enhancement. The additional knowledge and the design tools that will result from these research efforts will provide an optimized, safe, and conscious use of the resources for the built environment.

Acknowledgements Financial support from the European Union's Horizon 2020 research and innovation programme under the Marie Sklodowska-Curie grant agreement 'LIGHTEN - Ultralight membrane structures toward a sustainable environment' H2020-MSCA-ITN-2020-LIGHTEN-956547 is gratefully acknowledged.

References

1. Sustainable Development Goals from United Nation. https://www.un.org/sustainabledevelopment/. Accessed 3 Dec 2022.
2. International Energy Agency and the United Nations Environment Programme. (2018). 2018 Global Status Report: Towards a zero emission, efficient and resilient buildings and construction sector.
3. United Nations. (2021). 2021 Global Status Report for Buildings and Construction: Towards a Zero-emission, Efficient and Resilient Buildings and Construction Sector.
4. George Gastin, Forthbridge, Edinburgh, SCO. https://creativecommons.org/licenses/by-sa/3.0.
5. Felipe Gabaldón, Oceanographic, Valencia ES. https://creativecommons.org/licenses/by/2.0.
6. Drew, P. (1979). *Tensile architecture*. Routledge.
7. Seidel, M. (2009). *Tensile surface structures: A practical guide to cable and membrane construction*. Wiley.
8. Bischoff, M., Ramm, E., & Bieber, S. (2018). *Computational methods for Shell analysis*. IBB-Universität Stuttgart.
9. Hutchinson, J. W. (1966). A survey of some buckling problems, AIAA Jn14, 1505.

10. Barbero, E., & Tomblin, J. (1993). Euler buckling of thin-walled composite columns. *Thin-Walled Structures, 17*(4), 237–258.
11. Ju, G. T., & Kyriakides, S. (1992). Bifurcation and localization instabilities in cylindrical shells under bending—II. Predictions. *International Journal of Solids and Structures, 29*, 1143–1171.
12. Kyriakides, S., & Ju, T. (1992). Bifurcation and localization instabilities in cylindrical shells under bending Part I: Experiments. *International Journal of Solids and Structures, 29*, 1117–1142.
13. Deutsches Institut für Bautechnik, Pfeifer Seil und Hebetechnik Gmbh. (2018). European Technical Assessment ETA-11/0160 PFEIFER Wire Ropes.
14. Williams, C. J. K. (1990). Travelling waves and standing waves on fabric structures. *Structural Engineer, 68*(21), 432.
15. Healey, T. J., Bin, C. Q., & Li, R. (2013). Wrinkling behavior of highly stretched rectangular elastic films via parametric global bifurcation. *Journal of Nonlinear Science, 23*, 777–805.
16. O2 Arena from the Emirates Air Line cable car. cc-by-sa/2.0 – © PAUL FARMER – geograph.org.uk/p/3018490.
17. Osserman, R. (2010). Mathematics of the gateway arch. *Notices of the AMS, 57*(2), 220–229.
18. Meeks, W. H., & Perez, J. (2011). The classical theory of minimal surfaces. *Bulletin of the American Mathematical Society, 48*(3), 325–407.
19. Argyris, J. H., & Scharpf, D. W. (1972). Large deflection analysis of prestressed networks. *Journal of the Structural Division, 98*, 633–654.
20. Schek Heidelberg, H.-J. (1974). The force density method for form finding and computation of general networks. *Computer Methods in Applied Mechanics and Engineering, 3*(1), 115–134.
21. Otter, J. R. H., Cassell, A. C., Hobbs, R. E., & POISSON. (1966). Dynamic relaxation. *Proceedings of the Institution of Civil Engineers, 35*(4), 633–656.
22. Taylor, J. E. (1976). The structure of singularities in soap-bubble-like and soap-film-like minimal surfaces. *Annals of Mathematics, 103*(3), 489–539.
23. He, L., Li, Q., Gilbert, M., Shepherd, P., Rankine, C., Pritchard, T., & Reale, V. (2022). Optimization-driven conceptual design of truss structures in a parametric modelling environment. *Structure, 37*, 469–482.
24. Vanderplaats, G. N. (1982). Structural optimization-past, present, and future. *AIAA Journal, 20*(7), 992–1000.
25. Haftka, R. T., & Gürdal, Z. (2012). *Elements of structural optimization*. Springer Science & Business Media.
26. Hoyer, S., Sohl-Dickstein, J., & Greydanus, S. (2019). *Neural reparameterization improves structural optimization*. http://arxiv.org/abs/1909.04240.
27. Chandrasekhar, A., & Suresh, K. (2021). TOuNN: Topology optimization using neural networks. *Structural and Multidisciplinary Optimization, 63*(3), 1135–1149.
28. Shin, D., Cupertino, A., de Jong, M. H. J., Steeneken, P. G., Bessa, M. A., & Norte, R. A. (2022). Spiderweb Nanomechanical resonators via Bayesian optimization: Inspired by nature and guided by machine learning. *Advanced Materials, 34*, 2106248.
29. Kmet, S., Tomko, M., Soltys, R., Rovnak, M., & Demjan, I. (2019). Complex failure analysis of a cable-roofed stadium structure based on diagnostics and tests. *Engineering Fail Analysis, 103*, 443–461.
30. Hu, J., Chen, W., Zhao, B., & Yang, D. (2017). Buildings with ETFE foils: A review on material properties, architectural performance and structural behavior. *Construction and Building Materials, 131*, 411–422.
31. Ward, I. M., & Sweeney, J. (2012). *An introduction to the mechanical properties of solid polymers* (3rd ed.). Wiley.
32. De Focatiis, D. S. A., & Gubler, L. (2013). Uniaxial deformation and orientation of ethylene-tetrafluoroethylene films. *Polymer Testing, 32*, 1423–1435.
33. Galliot, C., & Luchsinger, R. H. (2011). Uniaxial and biaxial mechanical properties of ETFE foils. *Polymer Testing, 30*, 356–365.

34. Surholt, F., Uhlemann, J., & Stranghöner, N. (2022). Temperature and strain rate effects on the uniaxial tensile behaviour of ETFE foils. *Polymers*. https://doi.org/10.3390/polym14153156
35. Moritz, K. (2007). *ETFE-Folie als tragelement*. PhD dissertation, Technische Universtät Munich.
36. Zhang, Y., Zhang, Q., & Lv, H. (2012). Mechanical properties of polyvinylchloride-coated fabrics processed with precontraint® technology. *Journal of Reinforced Plastics and Composites, 31*, 1670–1684.
37. Zhang, Y., Zhang, Q., Zhou, C., & Zhou, Y. (2010). Mechanical properties of PTFE coated fabrics. *Journal of Reinforced Plastics and Composites, 29*, 3624–3630.
38. LIGHTEN MSCA Project. (2020). www.lighten-itn.eu; https://cordis.europa.eu/project/id/956547. Accessed 1 Dec 2022.
39. Meng, L., & Wu, M. (2016). Study on stress relaxation of membrane structures in the pre-stress state by considering viscoelastic properties of coated fabrics. *Thin-Walled Structures, 106*, 18–27.
40. Stranghöner, N., Uhlemann, J., Bilginoglu, F., Bletzinger, K.-U., Bögner-Balz, H., Corne, E., Gibson, N., Gosling, P., Houtman, R., Llorens, J., Malinowsky, M., Marion, J.-M., Mollaert, M., Nieger, M., Novati, G., Sahnoune, F., Siemens, P., Stimpfle, B., Tanev, V., & Thomas, J.-C. (2016). *Prospect for European guidance for the structural Design of Tensile Membrane Structures*. Science and Policy Report (SaP-Report).
41. Houtman, R. (2013). *TensiNet European design guide for tensile structures appendix 5: Design recommendations for ETFE foil structures. TensiNet Association.*
42. Cabello, A., & Bown, A. C. (2019). Using a nonlinear thermo-viscoelastic constitutive model for the design and analysis of ETFE structures. In *Proceedings of IASS annual symposia. Vol. 2019. No. 23*. International Association for Shell and Spatial Structures (IASS).
43. Bosi, F., & Pellegrino, S. (2017). Molecular based temperature and strain rate dependent yield criterion for anisotropic elastomeric thin films. *Polymer (Guildf), 125*, 144–153.
44. Brinson, H. F., & Brinson, L. C. (2008). *Polymer engineering science and viscoelasticity: An introduction*. Springer.
45. Bosi, F., & Pellegrino, S. (2018). Nonlinear thermomechanical response and constitutive modeling of viscoelastic polyethylene membranes. *Mechanics of Materials, 117*, 9–21.
46. Li, J., Kwok, K., & Pellegrino, S. (2016). Thermoviscoelastic models for polyethylene thin films. *Mechanics of Time-Dependent Materials, 20*, 13–43.
47. Argyris, J., Is, D., & da Silva, V. (1992). Constitutive modelling and computation of non-linear viscoelastic solids. Part II: Application to Orthotropic PVC-coated fabrics. *Computer Methods in Applied Mechanics and Engineering, 98*, 159–226.
48. Li, Y., & Wu, M. (2015). Uniaxial creep property and viscoelastic–plastic modelling of ethylene tetrafluoroethylene (ETFE) foil. *Mechanics of Time-Dependent Materials, 19*, 21–34.
49. Crawford, R. J., & Martin, P. (1998). *Plastics engineering* (3rd ed.). Elsevier Butterworth-Heinemann.
50. Robinson-Gayle, S., Kolokotroni, M., Cripps, A., & Tanno, S. (2001). ETFE foil cushions in roofs and atria. *Construction and Building Materials, 15*(7), 323–327.
51. Lamnatou, C., Moreno, A., Chemisana, D., Reitsma, F., & Clariá, F. (2018). Ethylene tetrafluoroethylene (ETFE) material: Critical issues and applications with emphasis on buildings. *Renewable and Sustainable Energy Reviews, 82*, 2186–2201.
52. Cremers, J. (2013). Energy issues and environmental impact of membrane and foil materials and structures-status quo and future outlook. In *Proceedings of the conference sb13, implementing sustainability—Barriers and chances, Munich, Germany* (pp. 24–26).
53. Maywald, C., & Riesser, F. (2016). Sustainability – The art of modern architecture. In *Procedia engineering* (pp. 238–248). Elsevier Ltd.
54. Finlay, T. C. R. (2021). The carbon footprint of long span structures: Review of the millennium dome and subsequent tensile systems. In *IASS annual symposium 2020/21 and the 7th international conference on spatial structures*.
55. Serge Ferrari. (2020). Corporate Social Responsibility Report 2020.

56. Polyloop startup. https://polyloop.fr/. Accessed 3 Dec 2022.
57. Vector Foiltec, Asahi. (2018). Environmental product declaration – Texlon®-System with Fluon® ETFE FILM.
58. Vector Foiltec, Nowofol Kunststoffprodukte, Dyneon. (2021). Environmental product declaration – Texlon® system.
59. Robinson, L. A. (2004). *Structural Opportunities of ETFE (ethylene tetra fluoro ethylene)*. MSc Thesis, Massachusset Institute of Technology.
60. Achintha, M. (2010). Sustainability of glass in construction. In *Sustainability of construction materials* (pp. 79–104). Elsevier.
61. Flor, J. F., Liu, X., Sun, Y., Beccarelli, P., Chilton, J., & Wu, Y. (2022). Switching daylight: Performance prediction of climate adaptive ETFE foil façades. *Building and Environment, 209*.
62. Poirazis, H., Kragh, M., & Hogg, C. (2009). Energy modelling of ETFE membranes in building applications. In *11th international IBPSA conference, Glasgow, Scotland* (p. 144).
63. Gobain, S. *Thermal performance of glazing systems*. https://www.saint-gobain-glass.co.uk/en-gb/glass-and-thermal-insulation. Accessed 3 Dec 2022.
64. Alongi, A., Angelotti, A., Rizzo, A., & Zanelli, A. (2021). Measuring the thermal resistance of double and triple layer pneumatic cushions for textile architectures. *Architectural Engineering and Design Management, 17*, 334–346.
65. Tian, G., Fan, Y., Wang, H., Peng, K., Zhang, X., & Zheng, H. (2020). Studies on the thermal environment and natural ventilation in the industrial building spaces enclosed by fabric membranes: A case study. *Journal of Building Engineering, 32*, 101651.
66. Suo, H., Angelotti, A., & Zanelli, A. (2015). Thermal-physical behavior and energy performance of air-supported membranes for sports halls: A comparison among traditional and advanced building envelopes. *Energy and Buildings, 109*, 35–46.
67. Vector Foiltec, Khan-Shatyr project. https://www.vector-foiltec.com/projects/khan-shatyr-entertainment-center-megatent/. Accessed 3 Dec 2022.

Chapter 3
Development of Sustainable Concrete Using Treated Bamboo Reinforcement

T. Vamsi Nagaraju and Alireza Bahrami

3.1 Introduction

The traditional and most common material used in civil infrastructures is concrete [1]. Although it is weak in tension, it is more potent in compression. As a result, steel bars are common reinforcement members, which provide the tensile strength for concrete. However, using steel as a reinforcing material has significant drawbacks, including increased cost and non-renewability [2]. In addition, steel production is a noticeable contributor to emissions of greenhouse gases. Thus, efforts are being undertaken by a few scientists and academicians to suggest a low-cost, eco-friendly alternative to steel by employing materials that are readily available locally. Several studies have examined using agriculture-based fibers as concrete reinforcement in this area. Jute, coir fiber, sisal, palm leaves, and bamboo are a few common natural fibers that have been investigated in the past [3]. However, despite the positive outcomes of most of these investigations, bamboo still has a distinct edge over other renewable reinforcing materials [4].

Fast-growing bamboo, which resembles wood, is a member of the Poaceae family of grasses. It takes just a couple of years to reach its peak strength and 5 years to reach perfection [5]. The maximum tensile strength of certain bamboo species is equal to the elastic modulus of mild steel. As a result, bamboo can withstand tension and compression loads like steel bars, while many other plant-based reinforcing materials cannot. Moreover, bamboo requires 50 times less energy to manufacture one cubic meter of material per unit of tension than steel [6]. These characteristics

T. V. Nagaraju
Department of Civil Engineering, S.R.K.R Engineering College, Bhimavaram, India

A. Bahrami (✉)
Department of Building Engineering, Energy Systems and Sustainability Science,
Faculty of Engineering and Sustainable Development, University of Gävle, Gävle, Sweden
e-mail: Alireza.Bahrami@hig.se

A. Bahrami (ed.), *Sustainable Structures and Buildings*,
https://doi.org/10.1007/978-3-031-46688-5_3

of bamboo have attracted the interest of numerous studies for use as concrete reinforcement [6, 7].

Amin et al. [8] conducted an experimental study to evaluate the feasibility of using bamboo as reinforcement in cement mortar as a woven mesh. The results demonstrated that adding bamboo mesh considerably improved the tensile, flexural, and impact strengths of mortar while also providing the ductility and toughness. To reduce water absorption by bamboo, Harelimana et al. [9] coated it with bitumen and sand before utilizing it as reinforcement in beam and column components.

Experimental evidence presented by Govindan et al. [10] showed that bamboo-reinforced concrete specimens exhibited four times higher ultimate load-carrying capacity compared to unreinforced concrete. However, the study also revealed that the bond interface between bamboo and cement mortar was weaker due to the smooth bamboo surfaces and moisture content, resulting in reduced tension-carrying capacity [11]. Ramesh et al. [12] provided a concise overview of the construction usage of bamboo as a tensile element in concrete members, including beams, slabs, panels, blocks, and components subjected to stress-strain.

Chithambaram and Kumar [13] reported bamboo-reinforced concrete panels to construct affordable houses in mountainous regions. Bitumen-treated strips and sand-blasted bamboo sheets were employed for constructing wall and roof components. A cement plaster with a uniform thickness of 10 mm on each face was applied to enhance the structure's strength against static and impact loads [6, 14]. Load tests were done to assess the performance of composite members better.

Himasree et al. [15] evaluated prefabricated bamboo-reinforced concrete wall panels as an alternative to clay and brick masonry for constructing low-cost homes in villages. Bui et al. [16] studied the durability of bamboo and mentioned that extensive exposure to wetting and drying cycles did not affect its tensile strength or Young's modulus, which are crucial requirements for any reinforcement members embedded in concrete sections. Mali and Datta [17] assessed the structural strength of bitumen-coated bamboo-reinforced concrete columns, while Qaiser et al. [18] examined the performance of bamboo-based concrete beams. Meanwhile, Schneider et al. [19] investigated the flexural load-carrying capacity of beams reinforced with bamboo using stirrups and bamboo as the primary reinforcements. Previous studies have indicated that the weak bond between bamboo strips and cement mortar is the main reason for member failure [20].

However, limited literature is available on the most effective treatment methods to enhance the interface contact between bamboo and cement mortar. Hence, the current chapter aims to strengthen the bond at the bamboo and cement concrete composite matrix. The efficiency of the treated bamboo behavior can be found through the performance of bamboo-reinforced concrete beams. Compressive and tensile tests were conducted on steel-reinforced and bamboo-reinforced concrete beams to compare their load-carrying capacity, deflection, ductility, stiffness, and energy absorption.

3.2 Possibility and Potential for Bamboo

In Asia, while India possesses a much greater bamboo reserve than China, the latter's exports surpass India's significantly [21]. However, there are several challenges that the Indian bamboo industry needs to address effectively, including lack of awareness, unregulated harvesting in extensive bamboo-bearing areas, absence of standardized bamboo farming policies, restrictions in transportation, inadequate knowledge regarding the suitable species selection based on the environment and cultivation age, limited market access, failure to meet global standardization requirements, poor quality management, and non-compliance with quality standards [22].

The total bamboo-covered area in India exceeds the estimated value of 8.96 million hectares, reaching 13.96 million hectares [23]. This substantial coverage indicates a remarkable increase in bamboo yield within India. Figure 3.1 [23] illustrates the state-wise distribution of bamboo production in India, visually representing the bamboo industry's productivity across the country.

Only a very few of the 160 species in India have DNA bar-coding and sequencing, which might be beneficial for their identification and research of structural patterns. Additional difficulties include the industry's preference for traditional furniture design and the strong demand for contemporary design but insufficient supply. Furthermore, the area where bamboo is grown is far from an enterprise that uses bamboo.

Additional difficulties include culm splits, bamboo drying that lowers the market value, culm weakening from too much moisture, and fungal invasion. For rural residents in many Indian States who depend on bamboo, plantation diversity of bamboo is particularly crucial. Forty-eight million households reside in grass, thatch, or bamboo-walled homes in India, according to data from the 2012 census [23]. More than 60% of the populace resides in homes with bamboo walls in regions like Arunachal Pradesh and Assam. Bamboo wall weaving methods may be found all over the world. However, other regions worldwide use bamboo infill walling

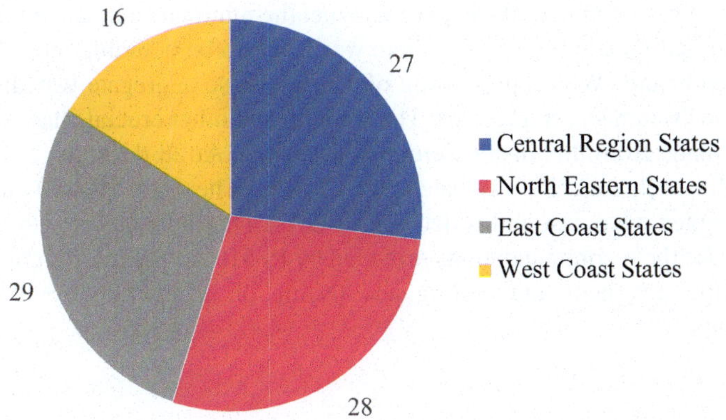

Fig. 3.1 Bamboo production percentage in India region-wise

systems with different building methods [23]. Bamboo is currently being considered as a building material that can aid in sustainable development [25]. Because of its advantages for the environment, society, and economy, it is exceptionally desirable.

3.3 Methodology

Bamboo was employed instead of steel in this experimental investigation on concrete beams. The locally accessible Muli Bamboo (*Melocanna bambusoides*) culms adhere to specific standards, such as being a maximum of 5 years old, representing a brownish look, and having been harvested before spring season. The lathe machine then cut bamboo strips into the required sizes from bamboo stems. Next, the bamboo sample's ultimate strength and elastic modulus were assessed using tensile testing. Next, the bamboo samples were treated twice for strengthening. First, bamboo samples were submerged in water for 24 h. The samples were then covered in a layer of lime. After that, the samples must dry for 30 days because immersion could cause them to absorb up to 32% of their weight in water. Afterward, these samples were given a bitumen coating on one set and an epoxy resin coating on the other. Finally, the samples were covered in a layer of dry sand to strengthen the binding between bamboo and cement mortar.

This study employed ordinary Portland cement (53 grade) with local market availability and standard requirements (IS 12269-1987) [25]. Sand and aggregate with the respective specific gravity values of 2.64 and 2.69 were utilized. The mixed design method described in IS 10262 and IS 456 was applied [25]. M20 grade concrete was used throughout the trial. Cement, fine aggregate, and coarse aggregate proportions for 1 m^3 of concrete were 365 kg, 670 kg, and 1250 kg, respectively. The water-cement ratio of the blends was kept constant at 0.5 for the concrete mix. Since cement was employed in the concrete mix, the bamboo-reinforced concrete members were intended for structural elements to be used in low-cost housing developments. Concrete of the experiment met the minimal cement content requirements and had no admixtures. Concrete was made to be pourable into cages made of bamboo beams. With a proportion of 70:30, coarse aggregate was divided into 20 mm and 10 mm size aggregates. The strength and other crucial characteristics of concrete made from this blend were also evaluated. At 28 days, the average cube strength for each concrete mix batch was 27 MPa. The 15 × 15 × 70 cm concrete prism samples were cast and tested for 28 days. The flexural strength was determined under the recommendations provided by IS 516-2004 [25]. Figure 3.2 shows the bamboo treatment and casting and testing of the bamboo-reinforced concrete beams.

Fig. 3.2 Bamboo-reinforced concrete beams: (**a**) bamboo treatment, (**b**) casting and testing of bamboo-reinforced concrete beams

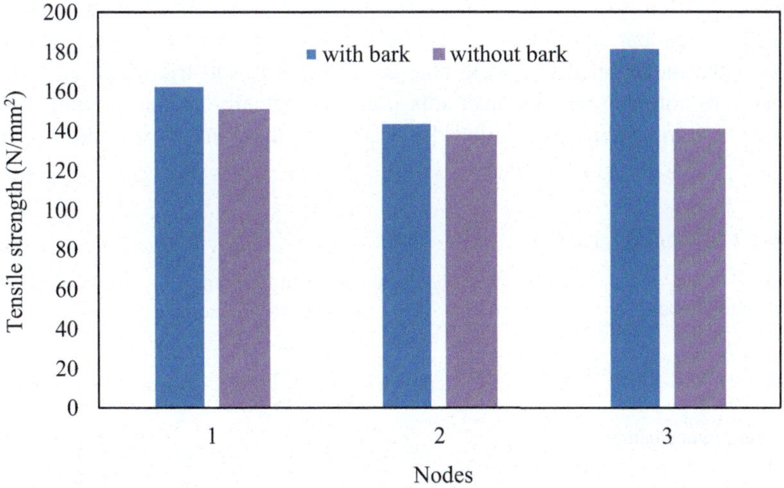

Fig. 3.3 Tensile strength of bamboo strips with and without bark

3.4 Results of Tested Bamboo-Reinforced Concrete Beams

To determine the bamboo reinforcement's ultimate strength and elasticity properties, tensile tests were performed on locally sourced bamboo strips. For this reason, a total of three samples were tested. Figure 3.3 displays the effect of nodes on the tensile strength of the bamboo strips with varying nodes.

According to the investigations, bamboo is more prone to node failure [6]. Three different places on the specimen were measured for breadth and thickness, and the average value was used to estimate the strength. The tests demonstrated that splitting caused samples to fail, most commonly at nodes. The bamboo strip had an average tensile strength of 171.22 N/mm^2.

3.4.1 *Results of Pullout Tests*

The effectiveness of different types of treated bamboo in increasing the interface strength between the treated bamboo and cement mortar was compared. Five cylinders were cast for comparison purposes: plain bamboo, bitumen-coated bamboo, epoxy resin-coated bamboo, bitumen-coated bamboo with sandblasting, and epoxy resin-coated bamboo with sandblasting.

At a depth of 100 mm, bamboo strips were put into the concrete cylinders. As per Agarwal et al. [6], the specimen's bond stress was computed. According to Table 3.1, the bitumen with sandblast-coated bamboo and epoxy resin with sandblast-coated bamboo concrete specimens exhibited the highest average bond strength. The bond force between bar deformations and the surrounding concrete was established through the mechanical interlock, chemical adhesion, and friction. However, it is essential to note that the chemical adhesion was typically less strong and could be easily overcome. Thus, it is often disregarded in the context of bond force. The

Table 3.1 Results of pullout tests

Type of treatment	Bamboo thickness (mm)	Bamboo width (mm)	Average bond stress (MPa)
Plain bamboo	31.24	4.26	0.141
	36.07	5.18	
	39.82	4.15	
Bitumen-treated bamboo	29.23	3.71	0.257
	34.12	3.56	
	32.15	4.16	
Epoxy resin-treated bamboo	37.91	5.12	0.314
	34.54	5.46	
	38.21	5.14	
Bitumen with sandblast-coated bamboo	42.14	3.85	0.518
	37.81	4.85	
	41.12	4.17	
Epoxy resin with sandblast-coated bamboo	33.45	5.22	0.572
	42.73	3.54	
	37.18	4.14	

remaining elements, namely the mechanical interlock and friction, work together to create a cohesive bond.

In analyzing the resulting stress, it is essential to distinguish between its longitudinal and radial components. The mechanical interlock mechanism is paramount as it is the critical factor in generating the binding force for distorted bars. Therefore, considering the mechanical interlock mechanism is crucial when studying this phenomenon.

In the case of other beam investigations, different materials were selected to explore their effectiveness. For instance, bitumen with sandblast-coated bamboo and epoxy resin with sandblast-coated bamboo concrete were chosen for their respective properties and potential benefits. Furthermore, these materials were specifically selected to assess their performance and suitability in the given context.

3.4.2 *Load-Deflection Behavior of Bamboo-Reinforced Concrete Beams*

All beams underwent full-scale testing, and the behavior was recorded as load-deflection curves. Table 3.2 lists the ultimate failure loads, ultimate failure moments, ultimate deflections, and first cracking loads of the beams.

Table 3.2 demonstrates that each bamboo-reinforced concrete specimen exhibited higher first crack loads and ultimate loads when compared to the control specimen. This was primarily due to the improved action of the bamboo-concrete composite, which produced a constant load distribution between the two components in contact. Also, the bamboo-reinforced concrete specimens with epoxy resin treatment had better flexural strength than those with bitumen coating. This was because the epoxy resin's stronger adhesive nature enhanced the friction between treated bamboo and concrete mortar.

Table 3.2 Load and deflection test results

Specimen	Ultimate failure load, P_u (kN)	Ultimate failure moment (kN.m)	Ultimate deflection (mm)	First cracking load (kN)
Steel-reinforced concrete beam (C)	59.12	14.71	22.63	9
Bitumen-coated bamboo-reinforced concrete beam with sandblasting (B1)	44.17	9.43	21.16	7
Epoxy resin-coated bamboo-reinforced concrete beam with sandblasting (B2)	51.28	12.24	19.23	11

Table 3.3 Ductility and stiffness behavior results

Specimen	P_i (kN)	P_u (kN)	Δx (mm)	Δy (mm)	Ductility index	Initial stiffness (kN/mm)	Final stiffness (kN/mm)
C	9	59	1.21	21.23	17.70	7.44	2.56
B1	7	40	2.91	19.65	7.98	2.63	1.68
B2	11	51	2.17	17.32	9.08	6.69	2.44

3.4.3 Ductility, Stiffness, and Energy Absorption of Bamboo-Reinforced Concrete Beams

The ductility, stiffness, and energy absorption of the beams are all important mechanical properties that influence how the beams react to stresses. Results for the initial first crack load (P_i), ultimate load (P_u), deflection at the first crack (Δx), deflection at the final crack (Δy), ductility index, initial stiffness, and final stiffness are presented in Table 3.3. The ductility index largely determines the mechanical properties of a material and its suitability for construction and infrastructure development. The ductility index of the bamboo-reinforced concrete beams was less than that of the control reinforced concrete beam (Table 3.3). The observed low tensile strength of the used bamboo could impact how ductile the concrete mix was. For other applications, such as low bearing buildings, where increased ductility is not necessary, this is not a big problem.

Stiffness is a term used to describe a material's resistance to deformation under load. In reinforced concrete beams, stiffness is frequently estimated as the slope of the load-deflection curves up to the yield point. Up to the first fracture load, Table 3.3 displayed that the control steel-reinforced concrete beam was more rigid than the bamboo-reinforced concrete beams. This illustrated how adding bamboo made concrete stiffer. For applications where stability and deflection control are crucial, a beam with high stiffness can support weights without exhibiting excessive deflection.

An energy absorption value can be calculated by dividing the area under the load-deflection curves up to the ultimate load by the area under the load-deflection curves up to the first crack load. Data on the energy absorption of the control steel-reinforced concrete and bamboo-reinforced concrete specimens are shown in Fig. 3.4. In comparison to the control beam, the bamboo-reinforced concrete beams (B1 and B2) exhibited higher energy absorption. This might result from the treated bamboo with sandblasting having superior energy-dissipation characteristics and effective load transfer between reinforcing bamboo (sandblasting roughened the surface) and concrete.

Even though the treated bamboo-reinforced concrete beams (B1 and B2) had higher initial deflection than the control beam, they can still be an acceptable substitute for steel-reinforced concrete beams in some situations. However, carefully evaluating bamboo treatment's unique requirements and characteristics is necessary to improve its performance and longevity further.

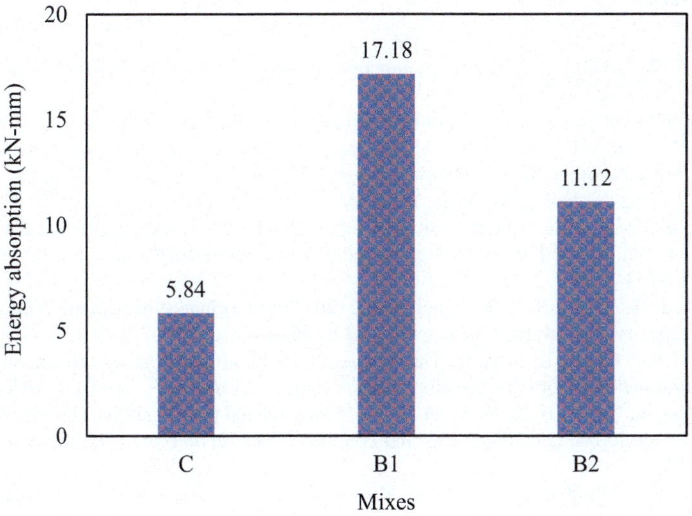

Fig. 3.4 Energy absorption of control steel-reinforced concrete and bamboo-reinforced concrete beams

3.5 Conclusions

To determine the possibility of employing bamboo as reinforcement in concrete, experimental research on various beams was conducted in this study. The tensile strength of bamboo strips with bark was higher than those without bark. Moreover, strips with two nodes were more durable than those with one or no nodes. Pullout tests revealed that, when compared to the other samples with adhesives alone, the sandblasted bamboo-reinforced concrete beams had higher interface resistance at the surface of the bamboo concrete composite. The bitumen-coated bamboo-reinforced concrete beams with sandblasting exhibited better energy absorption than the epoxy resin-coated bamboo-reinforced concrete beams with sandblasting or the control steel-reinforced concrete beam. The bamboo-reinforced concrete beams started to break between the loads 11 kN and 18 kN.

For temporary and lightweight residential houses, as well as in underdeveloped nations where steel for use as reinforcement members is either rare or too expensive, it is believed that bamboo may be utilized as a substitute for steel. However, reinforcement bars can only be replaced when more research is done on the best way to use the bars to maximize its potential as a composite-resistant material.

In the future, a total cost comparison of utilizing local versus conventional building materials for a specific region might also be beneficial. Research is required to determine how bamboo-reinforced concrete walls made of locally produced blocks operate in regions where bamboo is easily accessible and can be used as an alternative to steel.

References

1. Naik, T. R. (2008). Sustainability of concrete construction. *Practice Periodical on Structural Design and Construction, 13*(2), 98–103.
2. Thyavihalli Girijappa, Y. G., Mavinkere Rangappa, S., Parameswaranpillai, J., & Siengchin, S. (2019). Natural fibers as sustainable and renewable resource for development of eco-friendly composites: A comprehensive review. *Frontiers in Materials, 6*, 226.
3. Shah, I., Jing, L., Fei, Z. M., Yuan, Y. S., Farooq, M. U., & Kanjana, N. (2022). A review on chemical modification by using sodium hydroxide (NaOH) to investigate the mechanical properties of sisal, coir and hemp fiber reinforced concrete composites. *Journal of Natural Fibers, 19*(13), 5133–5151.
4. Moroz, J. G., Lissel, S. L., & Hagel, M. D. (2014). Performance of bamboo reinforced concrete masonry shear walls. *Construction and Building Materials, 61*, 125–137.
5. Li, Z., Chen, C., Mi, R., Gan, W., Dai, J., Jiao, M., et al. (2020). A strong, tough, and scalable structural material from fast-growing bamboo. *Advanced Materials, 32*(10), 1906308.
6. Agarwal, A., Nanda, B., & Maity, D. (2014). Experimental investigation on chemically treated bamboo reinforced concrete beams and columns. *Construction and Building Materials, 71*, 610–617.
7. Puri, V., Chakrabortty, P., Anand, S., & Majumdar, S. (2017). Bamboo reinforced prefabricated wall panels for low cost housing. *Journal of Building Engineering, 9*, 52–59.
8. Amin, K. F., Sharif, A., & Hoque, M. E. (2021). Bamboo/bamboo Fiber reinforced concrete composites and their applications in modern infrastructure. In *Bamboo Fiber composites: Processing, properties and applications* (pp. 271–297).
9. Harelimana, V., Zhu, J., Yuan, L., & Uwitonze, C. (2022). Investigating the bamboo as alternative partial replacement of steel bars in concrete reinforcement members. *The Structural Design of Tall and Special Buildings, 31*(6), e1921.
10. Govindan, B., Ramasamy, V., Panneerselvam, B., & Rajan, D. (2022). Performance assessment on bamboo reinforced concrete beams. *Innovative Infrastructure Solutions, 7*, 1–13.
11. Lima, H. C., Willrich, F. L., Barbosa, N. P., Rosa, M. A., & Cunha, B. S. (2008). Durability analysis of bamboo as concrete reinforcement. *Materials and Structures, 41*, 981–989.
12. Ramesh, M., Deepa, C., & Ravanan, A. (2021). Bamboo fiber reinforced concrete composites. In *Bamboo Fiber composites: Processing, properties and applications* (pp. 127–145).
13. Chithambaram, S. J., & Kumar, S. (2017). Flexural behaviour of bamboo based ferrocement slab panels with fly ash. *Construction and Building Materials, 134*, 641–648.
14. Aziz, M. A., & Ramaswamy, S. D. (1981). Bamboo technology for low cost constructions. In *Appropriate Technology in Civil Engineering: Proceedings of the conference held by the Institution of Civil Engineers, 14–16 April, 1980* (pp. 110–112). Thomas Telford Publishing.
15. Himasree, P. R., Ganesan, N., & Indira, P. V. (2023). Effect of opening on the performance of bamboo-reinforced Concrete Wall panels under one-way in-plane action. *Journal of Architectural Engineering, 29*(1), 04022036.
16. Bui, Q. B., Grillet, A. C., & Tran, H. D. (2017). A bamboo treatment procedure: Effects on the durability and mechanical performance. *Sustainability, 9*(9), 1444.
17. Mali, P. R., & Datta, D. (2020). Experimental evaluation of bamboo reinforced concrete beams. *Journal of Building Engineering, 28*, 101071.
18. Qaiser, S., Hameed, A., Alyousef, R., Aslam, F., & Alabduljabbar, H. (2020). Flexural strength improvement in bamboo reinforced concrete beams subjected to pure bending. *Journal of Building Engineering, 31*, 101289.
19. Schneider, N., Pang, W., & Gu, M. (2014, August). Application of bamboo for flexural and shear reinforcement in concrete beams. In *Structures congress 2014* (pp. 1025–1035).
20. Archila, H., Kaminski, S., Trujillo, D., Zea Escamilla, E., & Harries, K. A. (2018). Bamboo reinforced concrete: A critical review. *Materials and Structures, 51*, 1–18.
21. Rathour, R., Kumar, H., Prasad, K., Anerao, P., Kumar, M., Kapley, A., et al. (2022). Multifunctional applications of bamboo crop beyond environmental management: An Indian prospective. *Bioengineered, 13*(4), 8893–8914.

22. Dwivedi, A. K., Kumar, A., Baredar, P., & Prakash, O. (2019). Bamboo as a complementary crop to address climate change and livelihoods–insights from India. *Forest Policy and Economics, 102*, 66–74.
23. Parikh, N., Modi, A., & Desai, M. (2016). Bamboo: A sustainable and low–cost housing material for India. *International Journal of Engineering Research and Technology (IJERT)*, ISSN, 2278-0181.
24. Manandhar, R., Kim, J. H., & Kim, J. T. (2019). Environmental, social and economic sustainability of bamboo and bamboo-based construction materials in buildings. *Journal of Asian Architecture and Building Engineering, 18*(2), 49–59.
25. Reddy, A. N., & Meena, T. (2017, November). An experimental investigation on mechanical behaviour of eco-friendly concrete. In *IOP conference series: Materials science and engineering* (Vol. 263, No. 3, p. 032010). IOP Publishing.

Chapter 4
Implementation of Circular Economy Between Mining and Construction Sectors: A Promising Route to Achieve Sustainable Development Goals

Aiman El Machi and Rachid Hakkou

4.1 Introduction

For the past 150 years, the one-sense model of production and consumption has dominated the world. In this concept of a one-way supply chain, commodities undergo industrial processes to transform raw materials into finished products that are sold, utilized, and eventually disposed of either by incineration or as waste when they reach the end of their useful life. The raw materials that were initially gathered from nature are typically thrown away after a certain product has been consumed. This model is sometimes referred to as a linear model because it simply follows a linear route. It is more obvious that a linear business model cannot be continued for sustainable development due to the increasing growth in population, urbanization, and industrialization. Thus, growing demands for resource consumption lead to negative environmental impacts. However, policymakers, scholars, global corporations, and implementers are more interested in the move from the current linear economic model to a circular one.

The Ellen MacArthur Foundation (EMF) released research at the World Economic Forum in 2012 that assesses the possible advantages of the shift to a circular economy (CE). Only a portion of the EU industrial sectors might potentially benefit from approximately US$630 billion annually [1]. CE has considerable environmental and social benefits in addition to its enormous economic advantages, according to the EMF. To minimize resource leakage, CE is able to separate

A. El Machi (✉)
Faculty of Science and Technology, Cadi Ayyad University, Marrakech, Morocco
e-mail: aiman.elmachi@ced.uca.ma

R. Hakkou
Faculty of Science and Technology, Cadi Ayyad University, Marrakech, Morocco

Mohammed VI Polytechnic University, Ben Guerir, Morocco

A. Bahrami (ed.), *Sustainable Structures and Buildings*,
https://doi.org/10.1007/978-3-031-46688-5_4

economic growth from the use of resources. This can be done through the share, lease, reuse, repair, renovate, and recycle in a closed loop.

Contemporary society cannot function without metals. The demand for raw resources has expanded globally in recent decades, and between 2010 and 2030, resource utilization is predicted to even double [2]. The UN sustainable development objectives and the execution of the Paris Agreement adopted in 2015 led to the extensive use of a diversity of minerals for low-carbon applications.

As stated in Principle 8 of the International Council on Mining and Metals principles (ICMM), mining operations are linked to the use and disposal of goods that contain metals, involving the mining sector in efforts to close material cycles through recycling on a global scale. However, it is important to investigate potential at the mining site level before expanding the extractive industry's accountability across the full value chain. Otherwise, the construction sector is an important pillar of modern civilization. However, it has a negative environmental effect but contributes to environmental protection through the concept of sustainable construction [3]. Traditional construction gives little consideration to the effects on the environment and instead prioritizes cost, time, and quality. Nevertheless, sustainable construction tends to social demands while taking precautions to reduce any potentially harmful effects on the environment. Adopting sustainable construction practices can effectively lower the negative effects of construction activities.

It is very important to find adequate ways to reach the desired economic growth without negatively impacting the environment. This is only possible through the CE concept implementations in different sectors, and in diverse cycles (upcycle, downcycle). This strategy is a big enabler for both the mining and construction sectors to thrive through sustainability. In this chapter, the impact of the implementation of CE is discussed between the mining and construction industries. Case studies from several countries and companies are used as examples to provide a better vision. Moreover, multiple technical advances and future perspectives are presented to give more insight.

4.2 Mine Waste Management and Environmental Issues

4.2.1 Mining Industry and Waste Generation

The mining sector is an essential aspect in economic growth by providing vital raw materials and energy for many industries. However, its activities are still considered an impediment to natural resources, with negative environmental effects. In this context, the last decade has seen a renewed discussion about mining and sustainability due to public worries about the current degradation of the environment related to diverse mining processes.

Mine waste is the largest waste flow in the EU. Each year, mining activities produce huge quantities of mine waste. As stated by the Mining, Minerals, and Sustainable Development Project, about 3500 mineral waste facilities are active worldwide, composed mainly of dumps and dams. Over 100 billion tons of solid waste are produced annually from mining around the world presenting major environmental issues needing adequate management strategies [4].

The global estimated amount of waste from metal mining generated is approximately 4 gigatons/year [5]. At the same time over 1.2 billion tons of residue waste have previously been produced in the EU [6]. The valorization of mine waste is encouraged due to the financial and environmental expenses related to its management. In addition, these residues are considered raw materials in other technologies such as construction and buildings.

4.2.2 Environmental Risks Related to Mine Waste

Mining waste management involves the identification of diverse wastes and their effects on the environment and society in the framework of mining waste mitigation. The environmental risks of mining waste generation are acidity formation, heavy metals leaching, and stability of the tailings. Tailing dam catastrophes have happened in several countries. A dam is a storage space occupying several square kilometers of land, that could suffer from failures, responsible for environmental disasters. In 2008 when a dam accident happened in China 277 people died [7]. Tailings dam collapsed in 2015 in Brazil leading to the release of about 43.7 million m^2 of residues. In 2019, an iron tailings dam in Vale collapsed resulting in the death of 206 people [8]. Furthermore, waste rock and tailings cause severe human health complications resulting from the release of pollutants by leakage. Acid mine drainage (AMD) is a significant issue in mining activities. AMD is produced by sulfide oxidation in wastes releasing high content of sulfate, iron, and heavy metals characterized by acidic pH. AMD negatively impacts groundwater quality and limits the use of downstream water. The rehabilitation policies and the design of water treatment facilities remain important solutions to protect the surrounding environment. The remediation of AMD is still costly at around US$1.5 billion per year and the global environmental responsibilities are more than US$100 billion [9].

Different disposal methods of tailings are studied including conventional disposal; tailings reuse, recycling, reprocessing; and proactive management respecting the ICMM principle 8 (Table 4.1). The important statement is adopting an integrative approach for tailings management to improve environmental, social, and economic results. Thus, the recycling and reuse of mine tailings are new technologies to remediate environmental issues.

Table 4.1 ICMM principles

Principle number	Principle field of interest
1	Ethical business
2	Decision-making
3	Human rights
4	Risk management
5	Health and safety
6	Environmental performance
7	Conservation of biodiversity
8	Responsible design, use, reuse, recycling, and disposal of materials
9	Social performance
10	Stakeholder engagement

4.3 Valorization of Mine Waste for Construction Applications

4.3.1 Recycling of Mining Solid Waste for Construction Materials

In the framework of sustainable development goals (SDGs) of the UN and the implementation of the Paris Agreement, diverse minerals are used for green technologies considering low-carbon applications. The recycling of mine waste is quite compatible with the UN's SDGs. Mine tailing valorization participates notably in SDG11 "Sustainable Cities and Communities" and to SDG12 "Responsible Consumption and Production" by minimizing the output of waste. The management of mining waste currently consists of linear system thinking "take-make-waste." The employment of a CE model in the mining sector is necessary to resolve the issues of environmental pollution and to minimize waste with the generation of economic profits.

In another context, urbanization critically increases as 55% of the population lives in urban areas and will rise to 68% by 2050 implying a great consumption of raw materials to design building products [10]. To minimize this negative impact, the European Commission aimed to decrease the emissions related to the construction industry by 90% in 2050 [11]. Thus, the employment of residues to replace raw materials in construction may be an efficient method to attain CE goals by producing eco-friendly building materials with low embodied energy.

Mine wastes are considered a sustainable source of alternative materials in construction applications. These wastes are valorized as raw materials to produce clinker, bricks, aggregates, mortar, concrete, and geopolymers. The evaluation of the use of mine wastes and tailings as cementitious materials has roughly been investigated. Coal waste was used as raw material to produce clinker encouraging the saving of fossil fuels in kiln combustion [12]. Ceramics applications present an important alternative to reduce the costs associated with the management of mine tailings. In Russia, wastes from coal mines were recycled to produce ceramic bricks [13]. The fired bricks application is mainly targeted to allow the immobilization of the pollutant elements in the ceramic matrix [14].

Xu et al. [15] investigated the substitution of natural aggregates by iron tailings fine and coarse aggregates to produce concrete. Fontes et al. [16] used these tailings to produce ceramic tiles. Lam et al. [17] employed copper tailings as fine aggregates to elaborate concrete paving stones. More recently, Huynh et al. [18] tested the performance of fine-grained concrete samples using copper tailings. In addition, phosphate mine waste was recycled as coarse aggregates in concrete [19, 20]. Furthermore, Sedira et al. [21] recycled tungsten waste mud in alkali-activated binder. Furthermore, metallic tailings containing lead and zinc were valorized as aggregates for the formulation of mortar [22]. Mining waste from copper, gold, iron, bauxite, vanadium, zinc, and phosphate ores was used in the production of geopolymers for construction applications [23].

4.3.2 CE Concept Implementation in Mining Sector

The most recent approach to improving environmental sustainability is CE. Although Leontief [24] initially discussed CE in 1928, its first international application came when the German Parliament established a law on CE in 1996 [25]. Since the end of the 1990s, CE has been launched in China to solve environmental issues [26]. The visual representation of the CE concept is presented in Fig. 4.1. There is strong interest in CE in the political agenda in Europe to promote economic growth while reducing environmental impacts. In the UK, the implementation of a CE could produce 50,000 new jobs and attract investment of up to €12 billion [27]. The minerals and mining sector has made significant progress in the performance assessment of CE in relation to sustainable development, and the construction and building sector has also received substantial attention. Introducing a CE model into the mining sector could transform the industry into a sustainable one. Furthermore, the mining industry has enormous potential to adopt the CE approach by utilizing waste materials at different life cycle stages and developing eco-friendly products. Despite being in the early stages of the CE implementation, the development and implications of CE in the mining sector have grown significantly in recent years. The current CE agenda focuses on the supply chain of materials and the recovery of waste for direct reuse, which are two critical principles for a successful CE implementation.

The implementation of CE in the mining and construction/building sectors can significantly contribute to achieving the Sustainable Development Goal (SDG12), responsible consumption and production. By adopting a CE approach, these sectors can reduce waste and resource consumption while promoting resource efficiency and closed-loop systems. This can be achieved by recycling materials and products,

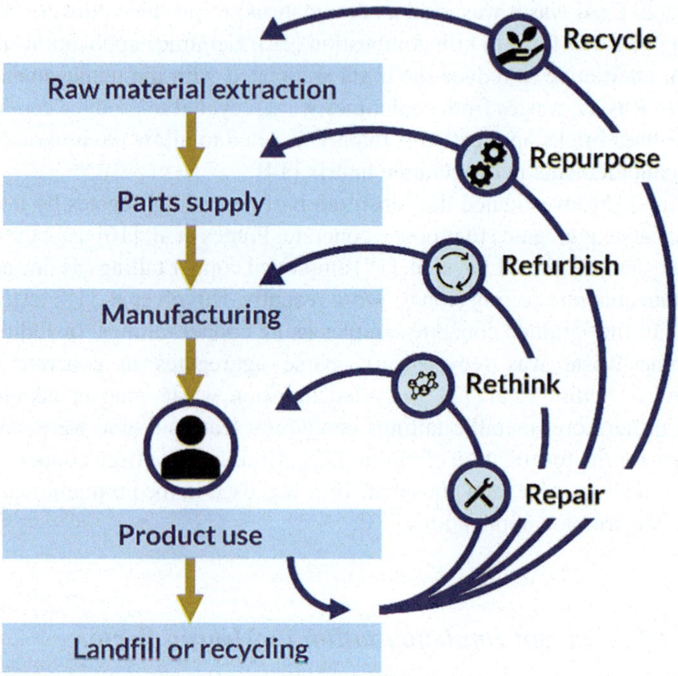

Fig. 4.1 Visual representation of CE concept

using renewable energy sources, and improving product design to extend their lifespan. In the mining sector, the use of waste materials as a source of secondary raw materials can reduce the need for extracting new materials from the earth, thus minimizing the environmental impact of mining operations. Similarly in construction, the use of recycled materials and prefabricated modular construction can minimize waste and resource consumption, while reducing construction time and cost.

Moreover, the implementation of CE can have positive effects on multiple SDGs. For instance, the use of renewable energy sources in mining and construction can contribute to achieving SDG7, affordable and clean energy. The development of closed-loop systems and resource-efficient technologies can promote SDG9, industry, innovation, and infrastructure, and SDG11, sustainable cities and communities, by reducing the environmental impact of industrial activities and improving the sustainability of urban areas. The reduction of water consumption and the treatment of wastewater can contribute to achieving SDG6, clean water, and sanitation, while the promotion of decent work and the creation of green jobs can promote SDG8, decent work, and economic growth. The reduction of greenhouse gas emissions and the adoption of sustainable land management practices can contribute to achieving SDG13, climate action, and SDG15, life on land, by mitigating the impact of industrial activities on the environment and preserving biodiversity. Figure 4.2 provides the percentage of addressed SDGs by projects related to the implementation of CE on the reuse of mine waste in the construction sector. A summary of the

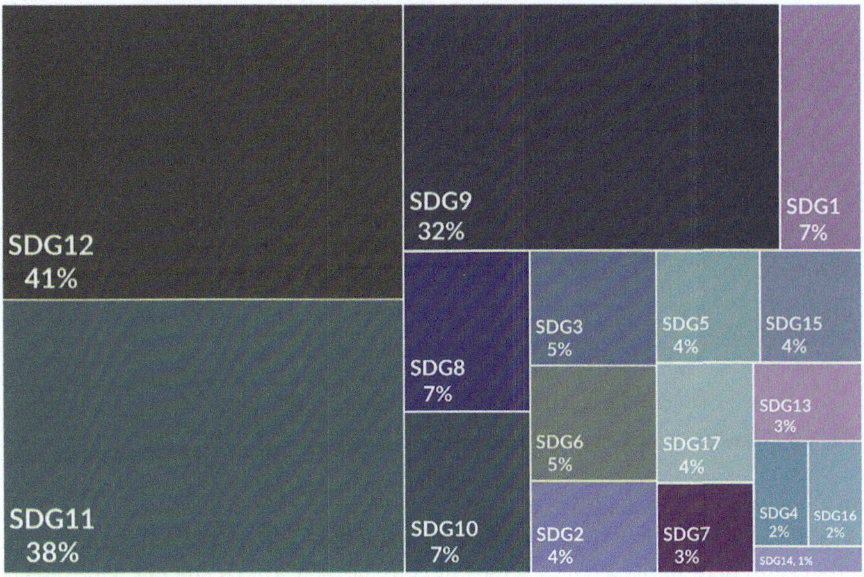

Fig. 4.2 Percentage of addressed SDGs by projects related to implementation of CE on reuse of mine wastes in construction sector

implementation status of CE in different countries shows that a global urge for such policy exists and some countries have reached a developed stage of implementation, while others remain at the beginning, as illustrated in Fig. 4.3. This fact is linked to many factors, the existence of laws, financial support and socio-economic awareness of the positive impact of adopting CE. The countries indicated in Fig. 4.3 are classified into four categories regarding their implementation of CE. Germany, Norway, South Korea, and the UK have advanced CE-driven societies and have attained a substantially greater level of the CE implementation. Australia, Canada, China, and the United States are in a less advanced state. Either they started the CE process more recently with significant implementation outcomes, or they started the CE process years ago with little success. Fewer countries, including Bhutan and Vietnam, have pioneered CE-driven societies through a number of actions and measures. In many areas of the economy, they began to minimize and reuse resources, and they saw progress. Afghanistan, Israel, and Morocco are at the beginning stages of implementing CE concepts, while there are a few isolated instances when resource recycling occurs on individual initiatives.

CE practices such as reusing and recycling mining waste have been successful in reducing resource waste. Anglo-American, a mining company based in South Africa, has implemented a zero-waste-to-landfill strategy by repurposing or recycling mining waste. To reduce the amount of hazardous waste sent to landfills, they employ a bioremediation plant that restores soil affected by hydrocarbon spills, thus promoting health, and protecting life on land, which is in line with SDG15 and SDG3. Recycling post-consumer minerals and metals at the end of their life through

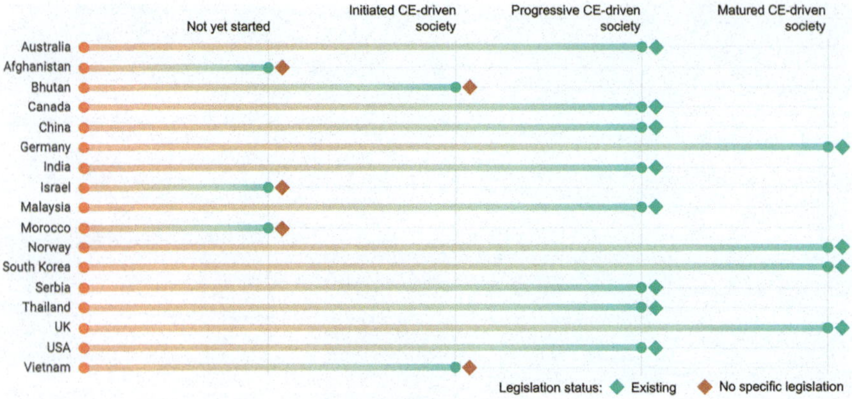

Fig. 4.3 Status of CE implementation in countries. (Data from [28])

mineral recovery can also give resources a new lease on life. For instance, Sumitomo Metal Mining was able to increase its copper scrap recovery rates significantly by processing high-purity copper scraps directly at their Tokyo smelter and refinery between 2010 and 2015. Lastly, by increasing comminution efficiency, energy recovery practices can recover the heat generated from mining processes and convert it into usable energy.

There are various global public policies that support circularity of minerals and metals. The CE Action Plan in Europe offers guidance and promotes best practices in mining waste management plans across the EU. The Canadian Minerals and Metals Plan emphasizes the continuous reduction of mining's environmental footprint by promoting CE practices, including transforming mine waste into useful products, enhanced mine closure planning, and environmentally reclaimed mine sites. The plan also includes systemic climate change adaptation planning. The Green Mining Initiative aims to improve energy efficiency, enhance productivity, and improve waste management and water management. The initiative has two main objectives: to reduce the environmental impacts of mining and to enhance Canada's competitiveness.

Mining in Australia has taken the lead in the development of CE in eco-industrial regions. Recently, regulatory strategies for metal recycling have been reevaluated based on the CE concept to revise regulations and their application to various stages of a metal's lifecycle from extraction to disposal. In Finland, Sitra's CE roadmap aims to reduce environmental impact, promote sustainable use of non-renewable resources, and utilize side streams. Mining waste including tailings, sludge, slag, dust, and ash can be considered secondary raw materials moving towards zero waste. In China, Urban Mining Pilot Cities aim to reduce the production of recyclable waste and promote sustainable urbanization. The CE model, consisting of "Reduce, Reuse, and Resource," was tested in six classes of UMPCs across 49 cities between 2010 and 2015 to improve industrial development and resource utilization. Morocco's mining industry holds 75% of the world's phosphate reserves but generates large quantities of waste. To reduce waste and generate new income, the phosphate mining company is working on recycling phosphate by-products for construction applications using various technologies.

The adoption of this approach by countries is an important index, but in parallel, the implementation of CE can be very effective while included in companies' business models and roadmaps. For instance, Boliden, a Swedish mining company, recovers metals from smelting slag and fly ash and turns them into construction materials. Canada-based Teck Resources recycles waste from its lead and zinc production to create new construction materials. Finnish start-up Betolar uses mining waste to create a low-carbon cement alternative, reducing carbon emissions from the cement industry.

Despite the progress made in implementing CE practices for mining waste reuse in the construction industry, there are still limitations and challenges that countries and companies face. One of the main challenges is the lack of standardization and certification for secondary raw materials. The quality and safety standards for these materials are not always clearly defined or consistent, which can create barriers to their use in construction. Another challenge is the transportation of mining waste to construction sites, which can result in increased greenhouse gas emissions and transportation costs. This issue can be mitigated by sourcing materials locally, but this may not always be possible. There are also financial challenges associated with implementing CE practices, as they may require upfront investment and long-term planning. Furthermore, regulatory frameworks and policies can vary greatly between countries and regions, which can create uncertainty for companies looking to implement CE practices. Finally, there may be social and cultural barriers to the adoption of CE practices, particularly in communities where mining has historically been a primary source of employment and economic development. It may take time and effort to build trust and support for new approaches to waste management and resource recovery.

4.3.3 Technical Advances and Future Perspectives

The reuse of mining waste in the construction industry is not a new concept. However, advances in technology have made it easier to transform these waste materials into useful products.

Advanced processing technologies including mechanical, thermal, and chemical processes are being used to transform mining waste into value-added products. These technologies are helping to develop new processes for transforming waste materials into valuable products and optimizing the design and construction of buildings using these materials. One example is the transformation of mining tailings into construction materials like lightweight aggregate, paving blocks, and bricks.

Geopolymer technology is a promising technique for the reuse of mining waste in the construction industry. This technique involves mining waste materials to produce geopolymer binders that could be used as a replacement for traditional cement in concrete production, reducing the environmental impact of cement production and providing a new market for mining waste materials.

Nanotechnology is being explored as a means of reusing mining waste materials in the construction industry. A process has been developed using nano-silica extracted from mining tailings to replace cement in concrete. This approach can improve the strength and durability of concrete while reducing its environmental impact.

The future perspectives for the implementation of CE to reuse mining wastes in the construction industry are promising. As the construction industry moves toward more sustainable practices, the reuse of mining waste materials can provide a valuable source of raw materials.

The integration of digital technologies like artificial intelligence, machine learning, and 3D printing is optimizing the reuse of mining waste materials in the construction industry. These technologies can help in developing new processes for transforming waste materials into valuable products and optimizing the design and construction of buildings using these materials. By doing so, the industry can make significant strides toward achieving sustainable and circular practices.

Circular business models are promising in promoting the reuse of mining waste materials in the construction industry. By creating new value chains for these materials, circular models create new revenue streams for the mining industry while reducing its environmental impact.

To assess the environmental impact of using mining waste materials in the construction industry, life cycle assessment (LCA) can be a useful tool. LCA can help identify the most sustainable use of these materials and optimize their use in the construction industry. LCA is also used to compare the environmental impact of using mining waste materials with that of traditional raw materials.

Material flow analysis (MFA) can assess the flows of materials within a system and identify opportunities for material reuse and recycling. MFA can be used to determine the sources and destinations of mining waste materials and optimize their use in the construction industry.

Blockchain technology can create a transparent and secure record of the provenance and quality of mining waste materials. This can increase their value and promote their reuse in the construction industry.

The Internet of Things (IoT) can also play a significant role in promoting the reuse of mining waste materials in the construction industry. By interconnecting physical devices and systems through the internet, IoT can enable data exchange and automation.

Digital twins are digital replicas of physical assets, systems, and processes. In the context of CE and mining waste reuse, digital twins can simulate and optimize construction processes, including the use of mining waste materials. By simulating the performance of construction materials made with mining waste under different environmental conditions, digital twins can identify optimal material formulations and manufacturing processes.

4.4 Conclusions

The world today faces environmental challenges growing over time and becoming an imminent threat to humankind. Industrialization is a huge step toward economic growth and well-being with a positive impact on humanity that cannot be neglected. But the negative impact on the environment should be emphasized at the same time. Thus, the need for a sort of agreement between the two is the only way to keep up the good part and tackle the problematic ones.

Mining operations generate significant amounts of waste. Their accumulation can cause dangerous environmental issues such as landslides, degradation of arable land, and acid mine drainage. Therefore, repurposing mining waste for practical purposes is crucial.

The construction industry is a major consumer of natural resources, causing significant environmental impact. An effective solution is to form partnerships between large waste rock producers and raw material consumers, which can reduce waste and raw material consumption while supporting economic growth and SDGs.

The CE implementation between those two big sectors is much more a need than a want today. This approach is the main directive to achieve the link between the above-mentioned environmental problems while giving room for the socio-economic thriving of businesses and countries. Accordingly, several countries are now starting their new way to achieving sustainability through the CE implementation. Technically, the CE implementation can only be done if the right techniques are involved. Therefore, LCA and other techniques are useful tools to assess the sustainability impacts of CE strategies.

Ultimately, this mixture of theoretical approaches and technical advances will never be possible to achieve without legal and administrative enabling. The urge in terms of administrative procedures and legislation is growing, and there should be a call to action in countries that are late in the CE implementation, after successful application experiences of the lead countries.

References

1. Ellen MacArthur Foundation. (2012). *Towards the circular economy: Economic business rationale for an accelerated transition*. Ellen MacArthur Foundation.
2. M. Reuter, C. Hudson, A. van Schaik, K. Heiskanen, C. Meskers, C. Hagelüken 2013 *Metal recycling: Opportunities, limits, infrastructure.*. A report of the working group on the global metal flows to the international resource panel.
3. Abidin, N. Z. (2009). *Sustainable construction in Malaysia developers' awareness*. http://eprints.usm.my/15580/1/Sustainable_Construction_In_Malaysia-Developers_Awareness.pdf. Accessed 2 Jan 2023.
4. Lèbre, É., Corder, G. D., & Golev, A. (2017). Sustainable practices in the management of mining waste: A focus on the mineral resource. *Minerals Engineering, 107*, 34–42. https://doi.org/10.1016/J.MINENG.2016.12.004

5. Haas, W., Krausmann, F., Wiedenhofer, D., & Heinz, M. (2015). How circular is the global economy? An assessment of material flows, waste production, and recycling in the European Union and the world in 2005. *Journal of Industrial Ecology, 19*, 765–777. https://doi.org/10.1111/JIEC.12244
6. Wang, C., Harbottle, D., Liu, Q., & Xu, Z. (2014). Current state of fine mineral tailings treatment: A critical review on theory and practice. *Minerals Engineering, 58*, 113–131. https://doi.org/10.1016/J.MINENG.2014.01.018
7. Ke, X., Zhou, X., Wang, X., Wang, T., Hou, H., & Zhou, M. (2016). Effect of tailings fineness on the pore structure development of cemented paste backfill. *Construction and Building Materials, 126*, 345–350. https://doi.org/10.1016/J.CONBUILDMAT.2016.09.052
8. Tayebi-Khorami, M., Edraki, M., Corder, G., & Golev, A. (2019). Re-thinking mining waste through an integrative approach led by circular economy aspirations. *Minerals, 9*, 286–289. https://doi.org/10.3390/MIN9050286
9. Manchisi, J., & Ndlovu, S. (2021). Prediction of acid mine drainage formation. In *Acid mine drainage* (pp. 31–56). https://doi.org/10.1201/9780429401985-3
10. Desa, U. N. (2018). *Revision of world urbanization prospects* (p. 16). UN Department of Economic and Social Affairs.
11. Report to Congress: Wastes from the Extraction and Beneficiation of Metallic Ores, Phosphate Rock, Asbestos, Overburden from Uranium Mining and Oil Shale | US EPA. (n.d.). https://www.epa.gov/hw/report-congress-wastes-extraction-and-beneficiation-metallic-ores-phosphate-rock-asbestos. Accessed 20 Dec 2022.
12. Malagón, B., Fernández, G., de Luis, J. M., & Rodríguez, R. (2020). Feasibility study on the utilization of coal mining waste for Portland clinker production. *Environmental Science and Pollution Research, 27*, 21–32. https://doi.org/10.1007/S11356-019-05150-W/TABLES/11
13. Stolboushkin, A. Y., Ivanov, A. I., & Fomina, O. A. (2016). Use of coal-mining and processing wastes in production of bricks and fuel for their burning. *Procedia Engineering, 150*, 1496–1502. https://doi.org/10.1016/J.PROENG.2016.07.089
14. Taha, Y., Benzaazoua, M., Hakkou, R., & Mansori, M. (2017). Coal mine wastes recycling for coal recovery and eco-friendly bricks production. *Minerals Engineering, 107*, 123–138. https://doi.org/10.1016/j.mineng.2016.09.001
15. Xu, F., Wang, S., Li, T., Liu, B., Li, B., & Zhou, Y. (2021). Mechanical properties and pore structure of recycled aggregate concrete made with iron ore tailings and polypropylene fibers. *Journal of Building Engineering, 33*, 101572. https://doi.org/10.1016/J.JOBE.2020.101572
16. Fontes, W. C., Franco de Carvalho, J. M., Andrade, L. C. R., Segadães, A. M., & Peixoto, R. A. F. (2019). Assessment of the use potential of iron ore tailings in the manufacture of ceramic tiles: From tailings-dams to "brown porcelain". *Construction and Building Materials, 206*, 111–121. https://doi.org/10.1016/J.CONBUILDMAT.2019.02.052
17. Lam, E. J., Zetola, V., Ramírez, Y., Montofré, Í. L., & Pereira, F. (2020). Making paving stones from copper mine tailings as aggregates. *International Journal of Environmental Research and Public Health, 17*, 2448. https://doi.org/10.3390/IJERPH17072448
18. Huynh, T. P., Nguyen, T. M., & Lam, T. K. (2022). Utilization of high volumes of copper mine tailings in the production of fine-grained concrete. *Materials Today Proceedings, 65*, 543–548. https://doi.org/10.1016/J.MATPR.2022.03.089
19. El Machi, A., Mabroum, S., Taha, Y., Tagnit-Hamou, A., Benzaazoua, M., & Hakkou, R. (2020). Valorization of phosphate mine waste rocks as aggregates for concrete. *Materials Today Proceedings*. https://doi.org/10.1016/j.matpr.2020.08.404
20. El Machi, A., Mabroum, S., Taha, Y., Tagnit-Hamou, A., Benzaazoua, M., & Hakkou, R. (2021). Use of flint from phosphate mine waste rocks as an alternative aggregates for concrete. *Construction and Building Materials, 271*, 121886. https://doi.org/10.1016/j.conbuildmat.2020.121886
21. Sedira, N., Gomes, J. C., & Luo, S. (2023). Characterisation and development of alkali-activated binders from co-utilisation of tungsten mining waste mud and electric arc furnace slag for precast applications. *SSRN Electronic Journal*. https://doi.org/10.2139/SSRN.4371201

22. Argane, R., Benzaazoua, M., Hakkou, R., & Bouamrane, A. (2016). A comparative study on the practical use of low sulfide base-metal tailings as aggregates for rendering and masonry mortars. *Journal of Cleaner Production, 112*, 914–925. https://doi.org/10.1016/J.JCLEPRO.2015.06.004
23. Mabroum, S., Moukannaa, S., el Machi, A., Taha, Y., Benzaazoua, M., & Hakkou, R. (2020). Mine wastes based geopolymers: A critical review. *Cleaner Engineering and Technology*, 100014. https://doi.org/10.1016/j.clet.2020.100014
24. Leontief, W. (1991). The economy as a circular flow. *Structural Change and Economic Dynamics, 2*, 181–212.
25. Barreiro-Gen, M., & Lozano, R. (2020). How circular is the circular economy? Analysing the implementation of circular economy in organisations. *Business Strategy and the Environment, 29*, 3484–3494. https://doi.org/10.1002/BSE.2590
26. Andersen, M. S. (2007). An introductory note on the environmental economics of the circular economy. *Sustainability Science, 2*, 133–140. https://doi.org/10.1007/S11625-006-0013-6
27. Hill, J. (2016). Circular economy and the policy landscape in the UK. In *Taking stock of industrial ecology* (pp. 265–274). Springer.
28. Ghosh, S. K. (2019). Circular economy: Global perspective. In *Circular economy: Global perspective* (pp. 1–452). Springer. https://doi.org/10.1007/978-981-15-1052-6/COVER

Chapter 5
Sustainable Geopolymer Bricks Manufacturing Using Rice Husk Ash: An Alternative to Fired Clay Bricks

T. Vamsi Nagaraju and Alireza Bahrami

5.1 Introduction

Bricks made of fired clay are widely utilized as construction materials around the globe, particularly for houses [1]. India is the main supplier of burnt clay bricks from South Asian nations. However, the most common methods for preparing bricks include cementing them or burning them in kilns at high temperatures between 800 °C and 1200 °C, which requires much energy [2]. Therefore, many chemical-activating wastes have been used as a constituent to lower the temperature required for brick production. Similarly, other by-products were added to cement to lower the amount of cement and improve its environmental sustainability [3]. Although fusing waste and substituting by-products for cement might assist production at lower temperatures, geopolymerization is a rather environmentally beneficial approach. The method utilizes less energy and emits less CO_2 since it requires a lower temperature [4].

Rich silica and alumina components are activated during the geopolymerization process in alkaline conditions [5]. Clay and ash are examples of precursor alumino-silicate materials. Fly ash, which is relatively rich in silica and alumina among the different precursors, is readily available in many regions [5, 6]. It has a wide range of applications in cementitious products, and numerous standards have been created to ensure its effective use. However, fly ash has some drawbacks; if the optimal range raises its amount in cement concrete, the consequent strength is diminished [7]. Fly ash can only be used at excessively high rates when activated in an alkaline

T. V. Nagaraju
Department of Civil Engineering, S.R.K.R Engineering College, Bhimavaram, India

A. Bahrami (✉)
Department of Building Engineering, Energy Systems and Sustainability Science, Faculty of Engineering and Sustainable Development, University of Gävle, Gävle, Sweden
e-mail: Alireza.Bahrami@hig.se

A. Bahrami (ed.), *Sustainable Structures and Buildings*,
https://doi.org/10.1007/978-3-031-46688-5_5

atmosphere. Many studies have also been done on fly ash-based geopolymers, but it is still difficult to make bricks by applying pressure during molding [8, 9]. Another readily available raw material is clay, which can be utilized as a precursor and an additional cementitious material. However, due to its high alumina and silica content, it may gain the strength through geopolymerization, which has a much lower temperature requirement than firing [10].

In an alkaline condition, precursors are activated, producing sufficient compressive strength. However, Bernal et al. [11] emphasized that the compressive strength exhibited noticeable differences based on the precursors' reactivity even when employing the same precursors. The particles' chemical properties and surface area, which differ according to the various sources of aluminosilicate minerals, all affect the precursor's reactivity [12, 13]. The primary consideration for choosing a precursor is the reaction rate depending on the presence of amorphous form of Si and Al in the raw ingredients to produce an excellent result [14]. Thus, the viability of a precursor must be evaluated early by assessing its level of reactivity. Fly ash is reactive, although clay is typically made reactive by alkaline activation or thermal treatment [15]. Clay is primarily cured in the 27–60 °C range after calcination. However, using such high temperatures to calcinate clay renders the entire activity undesirable; therefore, other eco-friendly methods should be used to eliminate such differences [16]. As a result, the temperature conditions need to be modified to change clay from a simple filler to a precursor for geopolymer [17].

Na/Al, Si/Al, molarity of NaOH, temperature condition, etc., are a few important factors that affect the compressive strength during geopolymerization [18]. When utilized together as an alkaline activator, NaOH and Na_2SiO_3 produce better strength growth than when used separately. Although primary studies maintained the alkali-activator ratio of 2.5 and further increased it to 3 to generate weak geopolymer, researchers found that Na_2SiO_3/NaOH equal to 1 produced a strong matrix for industrial fly ash and agricultural rice husk ash-based geopolymer blends [19, 20]. Low Ca fly ash-based geopolymer concrete produces greater compressive strength when the alkaline-to-precursor ratio ranges from 0.30 to 0.45 [21]. However, a semi-dry mixture that can be instantly de-molded is desired in brick. As a result, the alkaline to precursor ratio must be decreased, which affects the compressive strength and can be accommodated by molding pressure [22]. Molding pressure, particularly for geopolymerization, is another important factor that is sometimes overlooked, may favor the compressive strength. With a lower alkaline-to-precursor ratio, pressure makes precursors more wettable, which causes more considerable dissolution and increases the compressive strength [23].

This study presents the experimental investigation on how the percentage of rice husk ash, molarity of NaOH, and curing temperature affect the properties of geopolymer bricks. In addition, it uses leftover broken bricks and rice husk ash at the brick kiln as a precursor material to develop a new and sustainable brick for construction projects.

5.2 Methodology

This investigation used rice husk ash and brick waste powder as precursors, with sodium hydroxide pallets and sodium silicate gel. Waste bricks and rice husk ash were collected from the brick kiln site near Bhimavaram, India. Waste bricks were collected and crushed in an initial crusher unit before being ground into a fine powder that could pass through a 300 μm sieve. Figures 5.1 and 5.2 illustrate the raw materials gathered at locations in the field and X-ray diffraction patterns (XRD), respectively. The sodium silicate gel to sodium hydroxide solution ratio was kept at 1.5, and alkali materials included sodium hydroxide solutions with 3 and 5 molarities. Moreover, the alkali activator to solids ratio was kept at 0.45. Table 5.1 lists the chemical constituents of brick powder and rice husk ash.

In this research work, precursors were mixed for 5 min, an alkali-activated solution was added, and casting was done. A total of six mixes were prepared with varying percentages of rice husk ash as 0%, 10%, 20%, 30%, 40%, and 50% in the waste brick powder geopolymer blends. The reference mixes were designated as M0, M1, M2, M3, M4, and M5, respectively for 0%, 10%, 20%, 30%, 40%, and 50% of rice husk ash in the blends. The specimens were manually compressed using a wooden compressor to create bricks with the dimensions of 190 × 90 × 90 mm. To prevent the water evaporation, the specimens were then covered. The molds were removed after the specimens had been at room temperature for 24 h, and one set of specimens underwent typical ambient curing for 56 days, while the other set was cured in the oven at 100 °C for 24 h. This study examined the curing conditions and molarity of NaOH on the specimens. Research has revealed that the geopolymer mixes' CO_2 emissions, energy efficiency, and cost-effectiveness decreased as NaOH molarity was reduced [24].

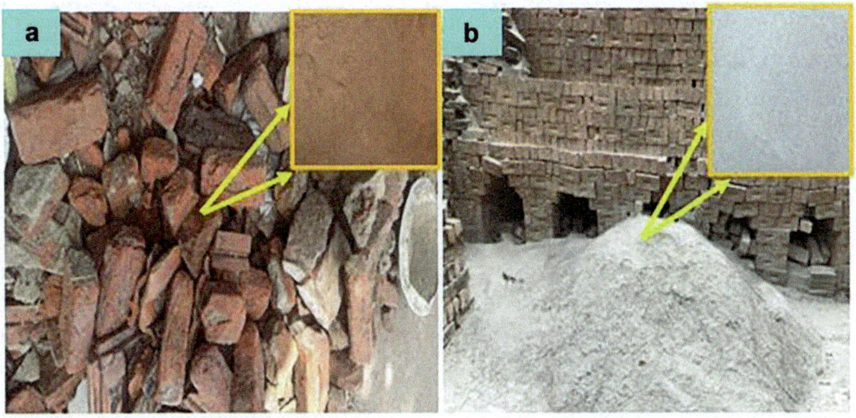

Fig. 5.1 Raw materials: (**a**) brick powder from waste bricks, (**b**) rice husk ash at brick kiln

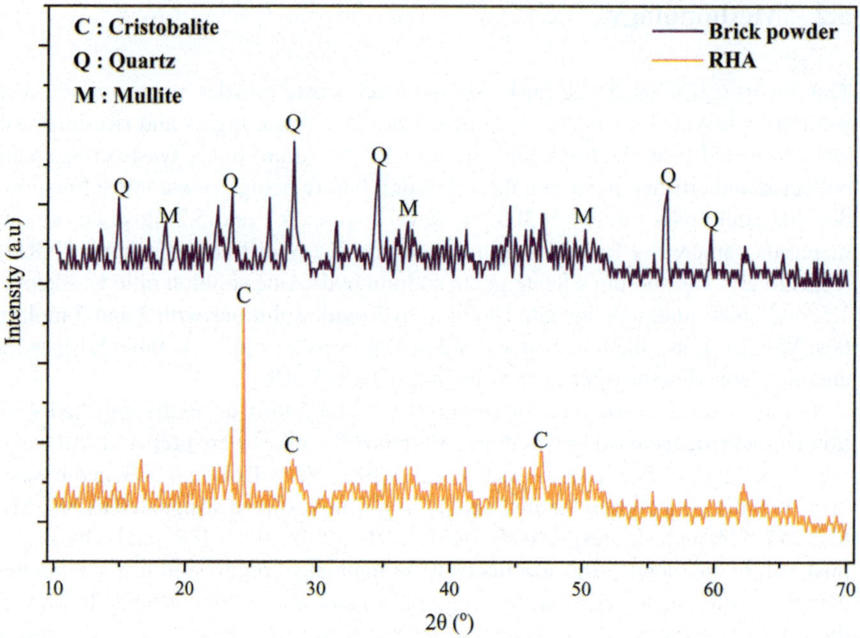

Fig. 5.2 XRD patterns of brick powder and rice husk ash (RHA)

Table 5.1 Chemical constituents of brick powder and rice husk ash

Materials	Chemical constituents (%)					
	SiO_2	Al_2O_3	MgO	K_2O	CaO	Fe_2O_3
Brick powder	69.4	13.6	1.7	4.1	1.3	7.4
Rice husk ash	95.4	0.4	0.5	0.1	0.7	0.6

5.3 Results and Discussion

5.3.1 Bulk Density and Strength Behavior of Geopolymer Bricks

Table 5.2 and Fig. 5.3 display the geopolymer bricks' bulk density and compressive strength, respectively, as a function of variations in molarity of NaOH and curing temperature. In the table and figure, 3M and 5M stand for 3 molarity and 5 molarity of NaOH, respectively. Molarity of NaOH and curing temperature were the crucial and significant parameters in geopolymerization reaction and from sustainability aspects. Si and Al can dissolve from the aluminosilicate source more readily when there is greater alkalinity.

In all mixes, bricks had a density between 1565 and 1810 kg/m^3. Moreover, a small increase in the density was noted in higher NaOH molarity of geopolymer bricks, since sodium hydroxide with 5 molarity was higher than 3 molarity. Higher

Table 5.2 Bulk density of geopolymer bricks

	Bulk density (kg/m³)			
	Ambient curing for 28 days		Oven curing at 100 °C for 24 h	
Mixes	3M	5M	3M	5M
MO	1745	1790	1780	1810
M1	1725	1745	1750	1760
M2	1675	1705	1690	1740
M3	1615	1670	1630	1690
M4	1590	1635	1605	1675
M5	1565	1580	1585	1640

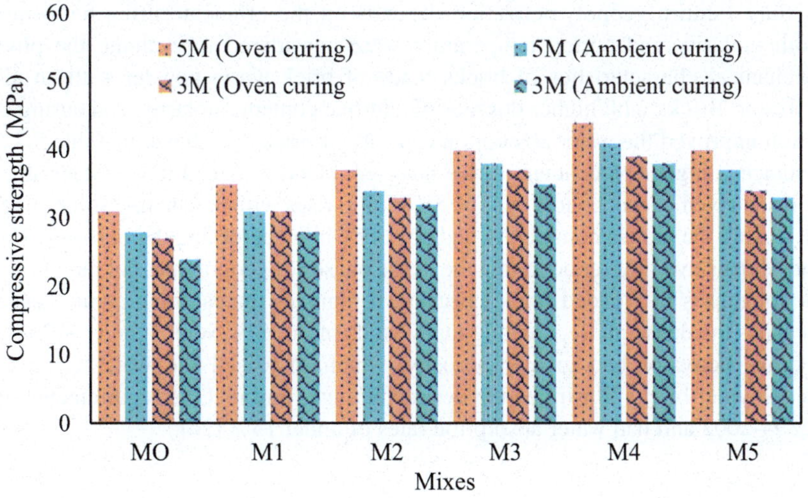

Fig. 5.3 Compressive strength of geopolymer bricks

compressive strength was obtained by high NaOH molarity and high NaOH molarity concentration. Compared to molarities of 3 and 5, an alkaline solution of 3 molarity demonstrated the maximum compressive strength in the present investigation. Since a lower NaOH molarity is less effective in the strength development, a similar result was observed in previous studies. All geopolymer bricks achieved higher compressive strength than the first-class brick standard as per IS 3495-1976, which is 105 kg/cm² (10.29 MPa) [25]. The highest compressive strength of 44 MPa (448.6 kg/cm²) was reached in the current study using 5 molarity alkaline solution and oven curing temperature of 100 °C for 24 h.

The compressive strength is divided into two categories by the Indian standard: the load-bearing range and the non-load-bearing range. A closer examination of clay brick waste-blended geopolymer bricks cast in this study and their compressive strengths in comparison to the standard indicated that all percentages of geopolymer bricks exhibited higher compressive strength than the standard load-bearing range (>5 MPa).

5.3.2 *Water Absorption Capacity of Geopolymer Bricks*

Table 5.3 depicts the geopolymer bricks' water absorption with varying NaOH solution concentration and curing condition. The Si/Al, Na/Al, and NaOH molarity are the important factors that affect the porosity of bricks, resulting in water absorption changes. The number of aluminosilicate bonds that forms increases with optimal molarity, Na_2SiO_3/NaOH ratio, and curing condition, making the mixture denser and more durable.

The higher percentages of rice husk ash improve the Al and Si leaching in the alkaline solution, and with higher concentration of NaOH the dense phases enhance, which results in fewer pores and impervious behavior of bricks. Additionally, the mixture's entire geopolymerization depends on the efficient curing temperature. Both molarity and curing temperature were considered to evaluate the physio-mechanical characteristics of bricks made of brick waste powder with an alkali activator. Bricks with higher dosages of additive content, molarity, and curing condition improved the water absorption capacity of bricks, as shown in Table 5.3. For instance, a higher percentage of rice husk ash blend provided lower water absorption at a given curing condition. The Si/Al ratio rises with an alkaline solution ratio, increasing the matrix complexity and density while decreasing the porosity. Maaze and Shrivastava [26] reported that the dense phases of geopolymer gel might impact the toughness of blended bricks, because it removes the essential component and weakens the dense phase, creating many voids in the mix. Some blends with a high alkaline activator had greater porosity, which increased the water absorption. Geopolymer bricks illustrated the desirable ranges and met the requirements of IS 12894-2002 and had water absorption rates of under 15% [26].

5.3.3 *Micro-Structural Behavior of Geopolymer Bricks*

Scanning electron microscopy (SEM) analysis was performed on the geopolymer brick mixes (M3 and M5), as displayed in Fig. 5.4. The surface texture and morphology of the brick specimens describe different aspects, such as the brick powder

Table 5.3 Water absorption of geopolymer bricks

	Water absorption (%)			
	Ambient curing for 28 days		Oven curing at 100 °C for 24 h	
Mixes	3M	5M	3M	5M
MO	14.5	13.9	13.8	13.1
M1	13.3	13.2	13.1	12.3
M2	12.7	12.2	12.4	11.4
M3	12.1	11.7	11.3	10.9
M4	13.7	12.5	11.7	12.1
M5	14.8	13.8	12.8	12.4

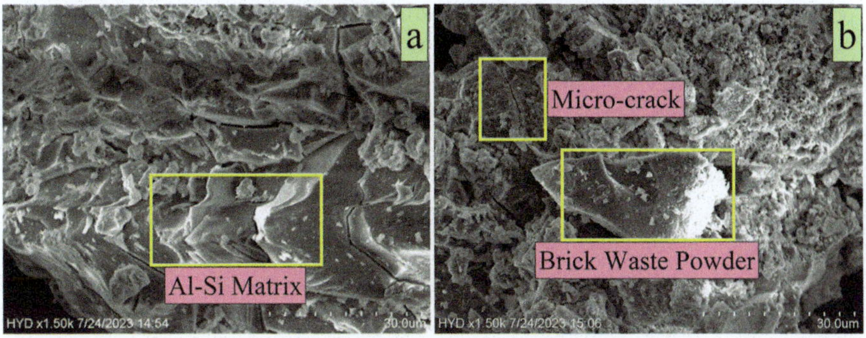

Fig. 5.4 SEM micrographs of mixes: (**a**) M3, (**b**) M5

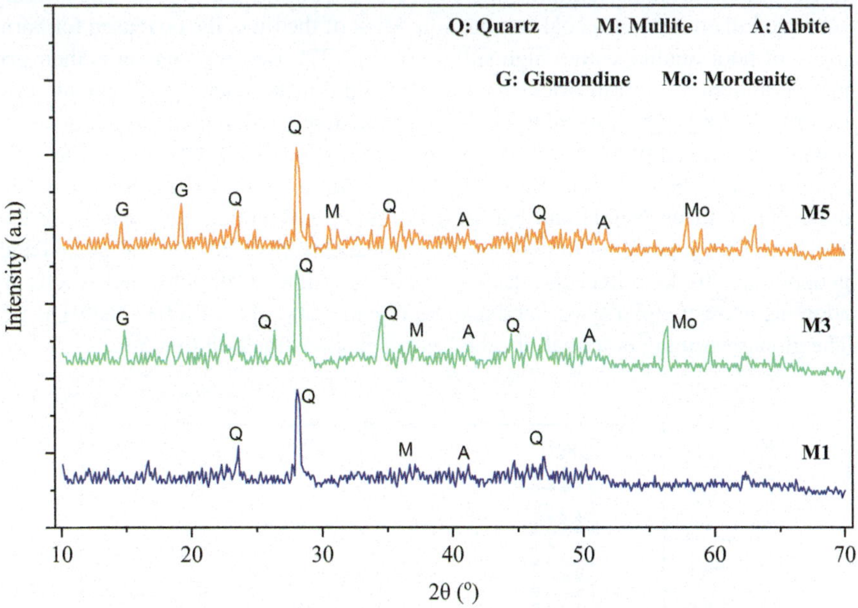

Fig. 5.5 XRD patterns of geopolymer bricks

and rice husk ash reactions in the geopolymer matrix. The SEM micrographs revealed the formation of many closely packed phases. Moreover, the amorphous silica presence of rice husk ash particles contributed to the Al-Si matrix. Developing a well-compacted geopolymer matrix could improve the porosity, water absorption, and compressive strength (Fig. 5.4).

Figure 5.5 depicts the XRD traces of geopolymer bricks with varying percentages of rice husk ash as replacement of brick powder, while the major peaks were observed in between the 2θ values of 22 and 34. The peaks represented the various crystal forms (quartz), and also indicated the other forms of crystalline particles present in the powder of waste brick. In addition, along with the albite ($NaAlSi_3O_8$), other traces such as orthoclase ($KAlSi_3O_8$) and gismondine ($CaAl_2Si_2O_8{\cdot}4H_2O$) were found in the geopolymer matrix. The tri-dimensional alumina silicate network (N—A—S—H) was found in geopolymer bricks in the form of mordenite ($Na_2Al_2Si_{10}O_{24}{\cdot}7H_2O$), which enhanced the dense phases and allowed the increase in the compressive strength and reduction in the water absorption.

5.3.4 *Sustainability Aspects of Geopolymer Bricks*

According to our findings, using clay bricks for walls has the most significant environmental impact because coal is utilized in the burning process. However, using bricks derived from agricultural waste has less environmental impact. The primary source of all emissions is coal combustion. Most of the time, the coal used for burning is of poor quality with a high sulfur content [27]. Because cement is their primary component, fly ash bricks have substantial effects. Each kg of cement emits roughly 0.83 kg CO_2 equivalent. Therefore, cement utilized in brick production contributes considerably to overall fly ash brick emissions [27]. Due to the lack of a firing process, geopolymer bricks from agricultural biomass blends have a less noticeable impact. NaOH and Na_2SiO_3 are important geopolymers with 1.88 and 1.915 kg CO_2 equivalent emissions per kg, respectively [28]. On the other hand, geopolymer bricks reflect the decreased consumption of Na_2SiO_3 and NaOH by adopting a lower molarity of 3 and maintaining the ratio of Na_2SiO_3 to NaOH at 1.5. The global warming potential of each brick is demonstrated in Fig. 5.6.

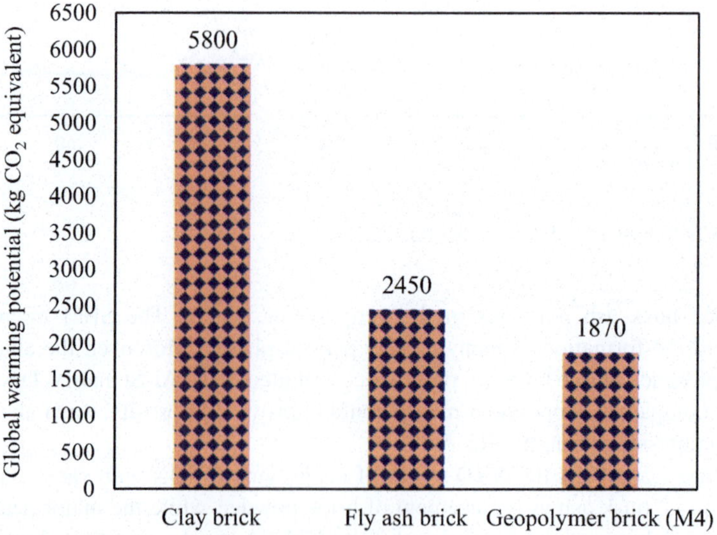

Fig. 5.6 Global warming potential of various bricks

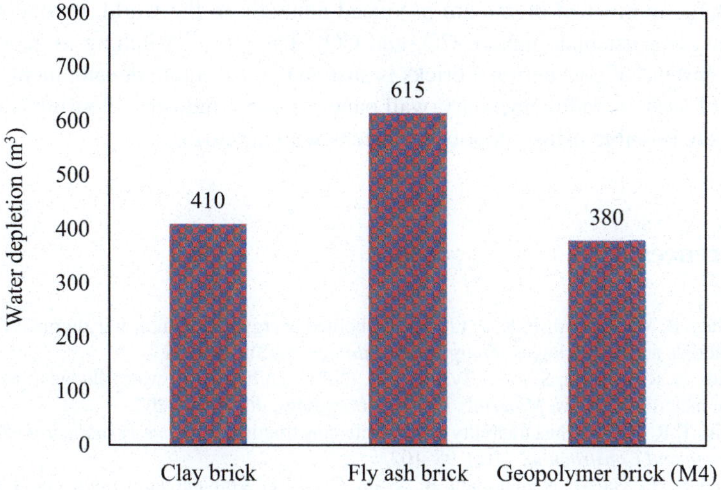

Fig. 5.7 Water depletion of various bricks

Figure 5.7 illustrates the operation's total water consumption, including brick and brickwork production. Since more water was needed to cure fly ash bricks, there was a considerable water shortage. Bricks made of clay and geopolymer had no cement content. Hence, water was not needed to cure them. The geopolymer brick's molding water content was lower than clay bricks.

5.4 Conclusions

This chapter focused on the possibilities for recycling waste bricks and agricultural waste rice husk ash to make construction materials made of geopolymers. The following conclusions can be drawn:

Addition of rice husk ash to geopolymer bricks decreased bulk densities due to lightweight (low specific gravity) rice husk ash. The highest bulk density was lower than 1700 kg/m^3, considerably lower than the range indicated in the standard (1700–2100 kg/m^3). It would therefore result in the production of lightweight and sustainable materials. When the amount of rice husk ash in geopolymer blends increased, the compressive strength enhanced dramatically. Furthermore, increased molarity and curing temperature showed stronger bonds at a given precursor concentration. The curing temperature and NaOH molarity concentration in the brick mixes had a substantial impact on the water absorption of the brick mixes. Increases in the dense matrix of the blends and consistent geopolymerization at higher curing temperatures minimized the water absorption of geopolymer brick specimens.

A large number of bricks are produced annually in the world, producing huge amounts of particulate matter, CO, and CO_2. Therefore, switching to geopolymer bricks instead of conventional bricks is sustainable for future development. A wide range of uses, including masonry, wall panels, pavers, industrial flooring, and canal lining are possible using geopolymer bricks with air curing.

References

1. Reddy, B. V., & Jagadish, K. S. (2003). Embodied energy of common and alternative building materials and technologies. *Energy and Buildings, 35*(2), 129–137.
2. Meda, S. R., Sharma, S. K., & Tyagi, G. D. (2021). Utilization of waste sludge as a construction material-a review. *Materials Today: Proceedings, 46*, 4195–4202.
3. Naik, T. R. (2008). Sustainability of concrete construction. *Practice Periodical on Structural Design and Construction, 13*(2), 98–103.
4. Bong, S. H., Xia, M., Nematollahi, B., & Shi, C. (2021). Ambient temperature cured 'just-add-water' geopolymer for 3D concrete printing applications. *Cement and Concrete Composites, 121*, 104060.
5. Kaur, M., Singh, J., & Kaur, M. (2018). Synthesis of fly ash based geopolymer mortar considering different concentrations and combinations of alkaline activator solution. *Ceramics International, 44*(2), 1534–1537.
6. Hajimohammadi, A., Provis, J. L., & van Deventer, J. S. (2011). The effect of silica availability on the mechanism of geopolymerisation. *Cement and Concrete Research, 41*(3), 210–216.
7. Bentz, D. P., Hansen, A. S., & Guynn, J. M. (2011). Optimization of cement and fly ash particle sizes to produce sustainable concretes. *Cement and Concrete Composites, 33*(8), 824–831.
8. Abdullah, M. M. A., Ibrahim, W. M. W., & Tahir, M. F. M. (2015). The properties and durability of fly ash-based geopolymeric masonry bricks. In *Eco-efficient masonry bricks and blocks* (pp. 273–287). Woodhead Publishing.
9. Bai, Y., Guo, W., Wang, J., Xu, Z., Wang, S., Zhao, Q., & Zhou, J. (2022). Geopolymer bricks prepared by MSWI fly ash and other solid wastes: Moulding pressure and curing method optimisation. *Chemosphere, 307*, 135987.
10. Kuranchie, F. A., Shukla, S. K., & Habibi, D. (2016). Utilisation of iron ore mine tailings for the production of geopolymer bricks. *International Journal of Mining, Reclamation and Environment, 30*(2), 92–114.
11. Bernal, S. A., Rodríguez, E. D., Mejia de Gutiérrez, R., Provis, J. L., & Delvasto, S. (2012). Activation of metakaolin/slag blends using alkaline solutions based on chemically modified silica fume and rice husk ash. *Waste and Biomass Valorization, 3*, 99–108.
12. He, J., Zhang, J., Yu, Y., & Zhang, G. (2012). The strength and microstructure of two geopolymers derived from metakaolin and red mud-fly ash admixture: A comparative study. *Construction and Building Materials, 30*, 80–91.
13. Gharzouni, A., Ouamara, L., Sobrados, I., & Rossignol, S. (2018). Alkali-activated materials from different aluminosilicate sources: Effect of aluminum and calcium availability. *Journal of Non-Crystalline Solids, 484*, 14–25.
14. Alhawat, M., Ashour, A., Yildirim, G., Aldemir, A., & Sahmaran, M. (2022). Properties of geopolymers sourced from construction and demolition waste: A review. *Journal of Building Engineering, 50*, 104104.
15. Ogundiran, M. B., & Kumar, S. (2016). Synthesis of fly ash-calcined clay geopolymers: Reactivity, mechanical strength, structural and microstructural characteristics. *Construction and Building Materials, 125*, 450–457.
16. He, J., Jie, Y., Zhang, J., Yu, Y., & Zhang, G. (2013). Synthesis and characterization of red mud and rice husk ash-based geopolymer composites. *Cement and Concrete Composites, 37*, 108–118.

17. Kaze, C. R., Alomayri, T., Hasan, A., Tome, S., Lecomte-Nana, G. L., Nemaleu, J. G. D., et al. (2020). Reaction kinetics and rheological behaviour of meta-halloysite based geopolymer cured at room temperature: Effect of thermal activation on physicochemical and microstructural properties. *Applied Clay Science, 196*, 105773.
18. Huo, W., Zhu, Z., Chen, W., Zhang, J., Kang, Z., Pu, S., & Wan, Y. (2021). Effect of synthesis parameters on the development of unconfined compressive strength of recycled waste concrete powder-based geopolymers. *Construction and Building Materials, 292*, 123264.
19. Nawaz, M., Heitor, A., & Sivakumar, M. (2020). Geopolymers in construction-recent developments. *Construction and Building Materials, 260*, 120472.
20. Jumaa, N. H., Ali, I. M., Nasr, M. S., & Falah, M. W. (2022). Strength and microstructural properties of binary and ternary blends in fly ash-based geopolymer concrete. *Case Studies in Construction Materials, 17*, e01317.
21. Nagaraju, T. V., Bahrami, A., Azab, M., & Naskar, S. (2023). Development of sustainable high performance Geopolymer concrete and mortar using agricultural biomass-a strength performance and sustainability analysis. *Frontiers in Materials, 10*. https://doi.org/10.3389/fmats.2023.1128095
22. Waheed, A., Azam, R., Riaz, M. R., & Zawam, M. (2022). Mechanical and durability properties of fly-ash cement sand composite bricks: An alternative to conventional burnt clay bricks. *Innovative Infrastructure Solutions, 7*, 1–12.
23. Ahmad, M., & Rashid, K. (2022). Novel approach to synthesize clay-based geopolymer brick: Optimizing molding pressure and precursors' proportioning. *Construction and Building Materials, 322*, 126472.
24. Verma, M., Dev, N., Rahman, I., Nigam, M., Ahmed, M., & Mallick, J. (2022). Geopolymer concrete: A material for sustainable development in Indian construction industries. *Crystals, 12*(4), 514.
25. Sreekeshava, K. S., Arunkumar, A. S., & Ravishankar, B. V. (2020). Experimental studies on polyester geo-fabric strengthened masonry elements. In *Advances in computer methods and Geomechanics: IACMAG symposium 2019 volume 1* (pp. 727–736). Springer Singapore.
26. Maaze, M. R., & Shrivastava, S. (2023). Design development of sustainable brick-waste geopolymer brick using full factorial design methodology. *Construction and Building Materials, 370*, 130655.
27. Joglekar, S. N., Kharkar, R. A., Mandavgane, S. A., & Kulkarni, B. D. (2018). Sustainability assessment of brick work for low-cost housing: A comparison between waste based bricks and burnt clay bricks. *Sustainable Cities and Society, 37*, 396–406.
28. Alsalman, A., Assi, L. N., Kareem, R. S., Carter, K., & Ziehl, P. (2021). Energy and CO_2 emission assessments of alkali-activated concrete and Ordinary Portland Cement concrete: A comparative analysis of different grades of concrete. *Cleaner Environmental Systems, 3*, 100047.

Chapter 6
Façade Fires in High-Rise Buildings: Challenges and Artificial Intelligence Solutions

Ankit Sharma, Tianhang Zhang, and Gaurav Dwivedi

6.1 Introduction

United Nations Member States adopted the 2030 Agenda for sustainable development in 2015, providing a blueprint for peace and prosperity for people and the planet [1]. Global partnerships are called for through its 17 sustainable development goals (SDGs), which are an urgent call for action by developed and developing countries. Goal 11 of these SDGs focuses on making cities and human settlements inclusive, safe, resilient, and sustainable [2]. To achieve this the construction industry is employing new techniques and materials to meet the ever-growing demand for high-rise buildings and make them sustainable. However, these high-rise buildings have associated fire safety risks that usually include [3–6]:

- Rapid external and internal spread of fire and smoke
- Difficult firefighting and rescue
- Difficult safe evacuation of the occupants
- Fire lasting for a longer time
- Stairwell filling with smoke
- Electrical short circuits

A. Sharma (✉)
Department of Mechanical and Aerospace Engineering, Case Western Reserve University, Cleveland, OH, USA

Department of Mechanical and Industrial Engineering, Indian Institute of Technology Roorkee, Roorkee, Uttarakhand, India

T. Zhang
Department of Building Environment and Energy Engineering, The Hong Kong Polytechnic University, Kowloon, Hong Kong

G. Dwivedi
Energy Center, MANIT Bhopal, Bhopal, India

A. Bahrami (ed.), *Sustainable Structures and Buildings*,
https://doi.org/10.1007/978-3-031-46688-5_6

- Cooking and heating equipment failure
- Carelessness and human error in handling and storing combustible materials

Façades are also being increasingly used to protect these buildings from wind, rain, and sunlight, and have become an integral part of the building. Figure 6.1 shows some of the buildings with façades.

The façade of a high-rise building is one of the components most susceptible to damage, especially in the event of a fire [7, 8]. Consideration of fire safety procedures has been hampered by visual appeal, efficiency of energy, materials, and sustainability. The problem has gotten more difficult as the number of high-rise structures has increased, providing a greater fire threat. The construction industry is utilizing various combustible materials in building façades to achieve sustainability and cost-effectiveness, which presents fire safety concerns [7, 9–11]. These are relatively low-frequency events, but they result in immense loss [12]. Many accidents worldwide have already proved the magnitude of fire risk involved in these façades (Fig. 6.2). Considering the severity of these accidents and to meet SDGs (specifically goal 11) of the Paris Agreement [1, 2], there is an urgent need to study façade fires to prevent future occurrence. However, due to their complex construction features and the presence of a variety of flammable items, fire scenarios can be complicated for these buildings. Undoubtedly, such fires cannot be reconstructed physically. Possible alternatives are assessing these complex fires in high-rise buildings using experimental set-ups and/or a modeling-driven approach [7, 11, 13–19] and developing test standards [20]. This chapter presents an overview of characteristics and testing methodology for understanding high-rise building fires with a particular focus on façades.

(a) (b)

Fig. 6.1 Commonly used façades in high-rise buildings (Source: Google Images)

Fig. 6.2 Grenfell tower fire accident in London UK (2017) [21]

6.2 Fire Growth: From Internal (Enclosure) Fires to External Spread

Fires are one of the most severe load situations for any construction. The capacity of the construction to endure then depends on how effectively it is planned. As previous big fires clearly have demonstrated, façade design may have a significant influence on the course and rate of fire spread. To understand how enclosure fires can spread to external façades, firstly there is a need to understand how fire initiates, grows, and propagates in a compartment and reasons for flashover occurrence. The advancement of fire spread is usually described as a curve dependent on time and energy or heat release rate (HRR) and is commonly known as fire growth curve, illustrated in Fig. 6.3. As can be seen, it consists of four main stages:

- Ignition and growth
- Flashover
- Fully developed fire
- Fire decay

Initially, the fire will start in a compartment if all three components of the fire triangle, i.e., fuel, heat, and oxygen, combine to produce a chain reaction. After the fire has ignited, its further growth is mainly influenced by the chemical and physical properties of fuel (heat of combustion, ignition temperature, chemical composition, density, etc.) and its configuration (shape and size) and hence also known as fuel-controlled fire. It is commonly expressed using the power-law equation as:

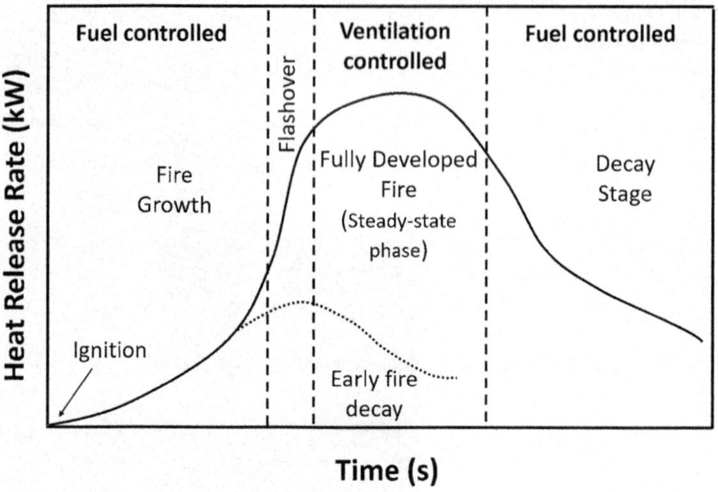

Fig. 6.3 Fire development curve for an enclosure [9, 22]

$$Q = \alpha t^2 \tag{6.1}$$

where Q (kW) is the heat release rate (HRR), t (s) is the time, and α (kW/s^2) is fire growth coefficient. HRR is not a fundamental property of fuel and has to be determined from testing either using oxygen consumption calorimetry (ASTM E1354) [23] or using mass burning rate $\dot{m}$ (kg/s) as:

$$Q = \dot{m}\Delta H_c \tag{6.2}$$

where ΔH_c is the heat of combustion (kJ/kg). HRR can be as low as 5 W for a burning cigarette [23] or can range from 7 kW to 2 MW for a liquid pool ($d = 0.1$–1 m) [24–26]. HRR growth period may be ultra-fast, fast, medium, or slow depending upon the critical time of 75, 150, 300, and 600 s, respectively, to reach 1055 kW [23]. It is characterized by the appearance of a fire plume that reaches the ceiling, and further spread occurs first as a ceiling jet and then in the form of a smoke layer. The short stage between growth and the fully developed fire is known as flashover when the fire spreads rapidly, and all the combustibles present in the compartment are involved. Flashover occurs when the temperature in the ceiling level has reached 600 °C or radiation to the floor of the compartment is greater than 20 kW/m^2 [23]. After flashover, fire reaches its fully developed state. It also marks a transition of fire from fuel-controlled to ventilation-controlled as at this stage, all the fuel is involved in the fire, so further growth will depend on the amount of air/oxygen available. After all the combustible items have burnt, fire again becomes fuel-controlled and starts decaying due to decreased fuel load and eventually extinguishes when no fuel is left.

It must be noted here that the fire curve depicted in Fig. 6.3 is ideal and need not be necessarily followed by all enclosures. It depends on various parameters, including ventilation conditions and available fuel load, which directly affect fire growth by changing the peak heat release rate and burning duration. If during the growth stage, sufficient ventilation and fuel are not available, fire may not reach flashover or fully developed stage and experience early decay without burning the entire compartment, as displayed in Fig. 6.3.

Enclosure fires often lead to external fires depending upon the energy released during the fully developed stage, which being ventilation-controlled is dependent on the ventilation factor (F_v) [27, 28] an expression derived for the ventilation factor as:

$$F_v = A_0 H_0^{0.5} \tag{6.3}$$

where A_0 (m^2) and H_0 (m) are the area and height of the ventilation opening, respectively. Based on this ventilation factor, maximum heat release rate HRR_{in} (kW) in fully developed stage inside an enclosure can be calculated as:

$$HRR_{\text{in}} = 1500 \ A_0 H_0^{0.5} \tag{6.4}$$

Enclosure fires generating HRR higher than HRR_{in} result in flames coming outside the doors and windows with heat release rate HRR_{ext} (kW) as:

$$HRR_{ext} = HRR - HRR_{\text{in}} \tag{6.5}$$

HRR_{ext} impinges directly on the façade, and in the case of high-rise buildings with combustible façades, it will lead to rapid vertical fire spread from one floor to another depending upon the type of façade system installed on the building. Commonly used façade systems are discussed in the next section.

6.3 Façade Systems and Types

Designed by advanced engineering methods and techniques, these external façade systems have become complex and constitute a significant percentage of the building's overall cost. In recent years, the configuration of façade systems has changed from simpler to complex to meet various needs like protection from wind, rain, and sunlight, providing insulation and aesthetic sense to the building. However, this shift to thinner, lighter, and more energy-efficient systems has also led to an increased potential risk of fires. Common types of façade systems demonstrated in Fig. 6.4 are described below [9, 12]:

(a) **Monolithic façades:** This is the simplest form of façade and consists of a single layer of non-combustible material such as concrete, brick, or glazed curtain

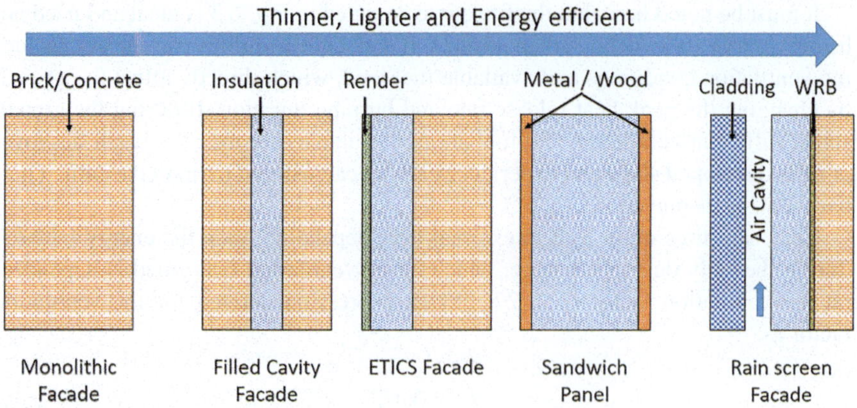

Fig. 6.4 Different types of façade systems [9, 12]

wall. Main fire hazards associated with these façades are shattering and spalling due to heat transfer from fire. However, they pose less risk in terms of flammability due to their non-combustible nature. Despite their lower flammability, they are not commonly used in modern infrastructures as they cannot meet energy efficiency requirements due to the presence of a single material only. Hence more complex systems, as described in the next sections, have been developed.

(b) **Insulated façades:** To fulfil the energy requirement, such façades are equipped with a layer of insulation material assembled between two layers of combustible material. Flammability of such façades depends on the choice of the material used for insulation. They are further divided into the following types depending upon the bonding material used for protecting insulation:

   (i) Filled cavity façades: They consist of insulation material filled between two layers of materials such as concrete or brick wall, which prevent direct impingement of flames on insulation. These are commonly used in low-rise buildings (<15–20 m).
   (ii) ETICS façades: This stands for External Thermal Insulation Composite Systems. They consist of insulation material between a thick layer of brick/concrete wall on the inner side and thin layer (2–12 mm) of render on the outer side for providing insulated and water-resistant finished surface. They provide enhanced thermal insulation and weather protection along with the improved exterior design of the building. However, due to a thin layer of protection, such systems are more prone to fire spread than filled cavity façades.
   (iii) Sandwich panels: In these systems, the insulation material is sandwiched between two thin layers of metals, plywood, or gypsum, for providing inert and aesthetic sense to façades. They are generally used as an external façade in high-rise buildings. Due to their aesthetics and affordability, they

are more commonly preferred by designers and architects. Similar to ETICS, they also pose fire threats if the inner core of insulation material is combustible.

(c) **Rainscreen façades**: In previously described façade systems, there was a common problem of moisture getting deposited over façades. To overcome this problem, rainscreen façades are developed where a small air cavity is kept for ventilation between the external façade and insulation. Generally, they are an assembly of mainly three components: exterior cladding with air cavity behind, continuous insulation (CI), and water/weather-resistive barriers (WRB), as depicted in Fig. 6.4.

(i) Exterior cladding and air cavity: It forms the outermost component of the assembly and is often known as rainscreen cladding as it is mainly designed to protect the buildings against rainwater. The primary types of exterior cladding materials used are aluminum composite panels (ACP), high-pressure laminates (HPL), and fiber-reinforced plastic (FRP). These materials can have a combustible, fire retardant, or mineral fill core, depending on the desired level of fire protection and cost. Various joint systems are used to install them, creating an air cavity of 25–100 mm behind the cladding. This air cavity serves two purposes: draining rainwater and promoting upward airflow within the cavity in hot weather. This ventilation helps to remove moisture from the façade, keeping it dry.

(ii) Continuous insulation (CI): The second element of the construction involves adding a layer of insulation to the exterior of the building structure to enhance the R-value of the outer wall. This improves energy efficiency by providing insulation. Common types of continuous insulation materials are expanded polystyrene (EPS), polyisocyanurate (PIR), phenolic foams, and mineral wool (MW). The thickness of the insulation depends on the climate zone and desired R-value. It is important to note that using combustible insulation in combination with combustible exterior cladding can exacerbate the situation in the event of a fire.

(iii) Weather/water-resistive barrier (WRB): The third element of the assembly is installed over the exterior sheathing and beneath the continuous insulation. This layer helps to prevent moisture damage to the building and regulates the relative humidity to maintain comfort inside the building. Building professionals commonly use either fluid-applied membranes or building wraps as water-resistive barriers (WRBs).

These façades are commonly used in high-rise buildings due to their high thermal efficiency, superior weatherproofing, and aesthetic sense. They also solve the problem of moisture control by having an air cavity. However, in the event of a fire, this internal cavity forms a chimney effect along with combustible façade causing rapid vertical fire propagation to other floors depending upon the fire scenarios as discussed in the next section.

6.4 Design Fire Scenarios for Façades – Internal and External Fires

Multiple fire scenarios can initiate façade fires, which can be divided into the following two main categories (Fig. 6.5):

6.4.1 Internal Fires

Internal fires inside the compartment of a building can occur due to different ignition/heat sources varying in size, intensity, and duration ranging from a tiny spark to flammable liquid fires. In modern infrastructures, there are a variety of factors that have led to frequent internal fires such as complex designs, open spaces (less compartmentation), increasing fuel loads, void spaces, and changing building materials. This has led to shorter time to flashover, faster fire propagation, shorter escape times, and increased exposure. It was also confirmed in an experiment conducted by UL (Underwriters Laboratories) where flashover time of less than 5 mins was observed in modern rooms as compared to 30 mins in legacy (old) rooms. These internal fires (pre-flashover or post-flashover) are among the most common ignition scenarios for façade fires when they come out due to flashover leading to breaking of windows. They can ignite combustible façades by direct impingement on the façade or penetrating the air cavity or void spaces as shown in Fig. 6.6 leading to more extensive flame propagation.

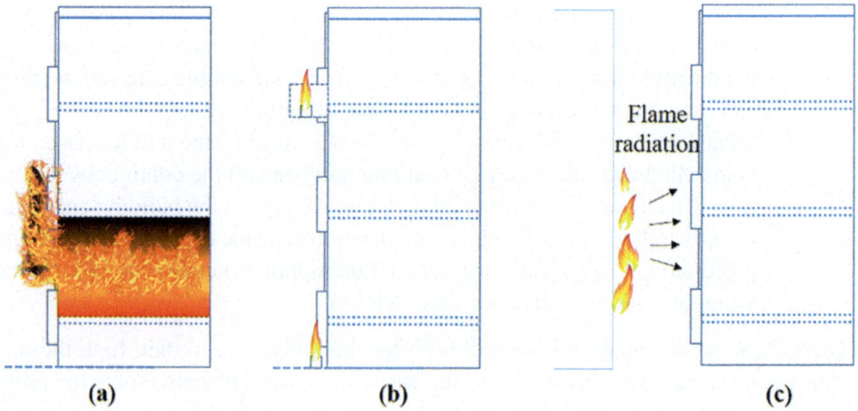

Fig. 6.5 Façade fire scenarios for high-rise building fire accidents [9, 29]

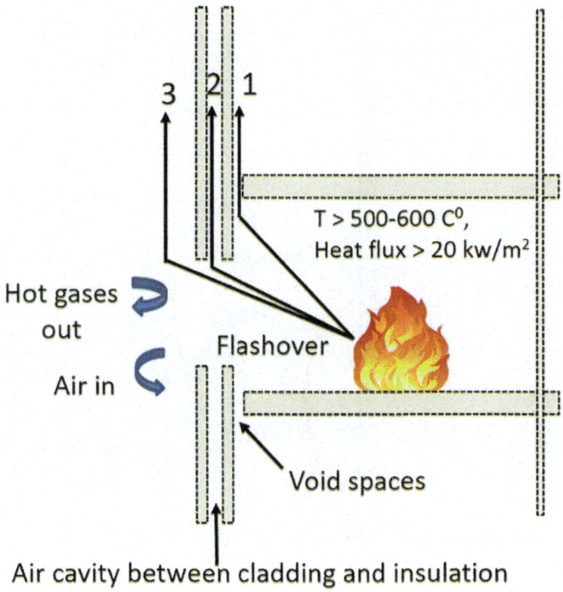

Fig. 6.6 Internal fires spreading via: (1) Void spaces, (2) Air cavity, (3) Direct impingement on exterior façade [9, 30]

6.4.2 External Fires

External fires near windows or balconies can directly impinge on the outer façade without the need for flashover to occur. Nearby burning buildings or combustible items (such as a parked car or trash cans) can also initiate façade fires by heating the façades via radiation.

Two major factors that can lead to the above fire scenarios are the use of combustible materials in the façade system (cladding, insulation, joints, etc.) and the absence of or insufficient vertical cavity barriers and fire stops to avoid fire penetration into the cavity leading to chimney effect as discussed in the next section.

6.5 Vertical Fire Propagation Mechanism over Façades

The flame spread can be defined as the process in which flame moves over a combustible pyrolyzing surface acting as fuel. The moving flame front can be considered as an invisible boundary between the unburnt and burnt part of combustible material. Flame front moves by heating unburnt fuel in the preheating zone from the flame in the pyrolysis zone, where actual burning and thermal degradation of the material occurs. How fast the fire grows will greatly depend on the heat transfer and flame spread rate. Hence, it is crucial to study the mechanism of flame spread and various factors affecting its rate. The rate of vertical fire spread has been noted to increase exponentially over time due to a doubling effect [31, 32].

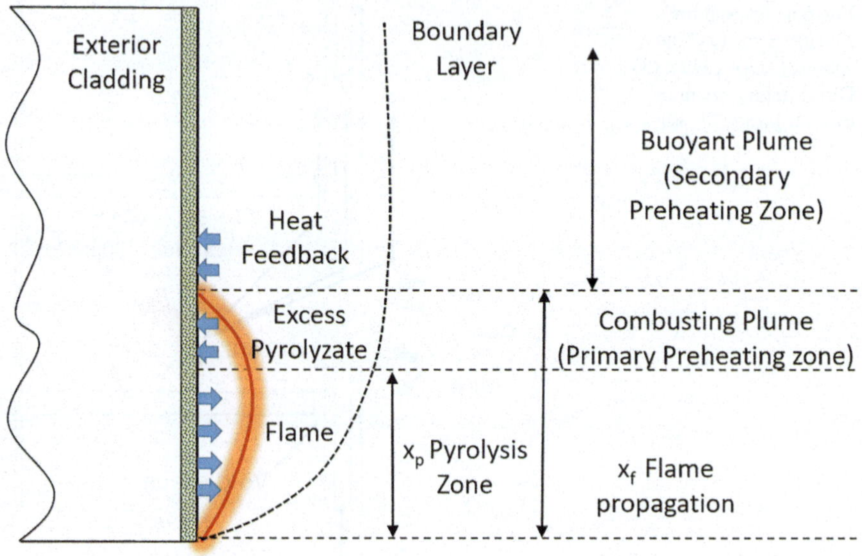

Fig. 6.7 Vertical upward fire spread mechanism over a façade [9, 33]

As shown in Fig. 6.7, three primary regions are observed in upward flame spread over façades. The first zone is the pyrolysis zone where the actual burning of material produces combustible volatiles that burn to produce a visible flame. The next region between x_p (pyrolysis zone) and x_f (flame propagation length) is known as a combusting plume where unburnt volatiles move upward due to buoyancy, burn, and also causes preheating of the unburnt solid fuel above the pyrolysis zone. The third, uppermost zone above x_f consisting of hot combustion products is known as a buoyant plume, where the physical flame is not present. Rate of flame spread over façades is mainly dependent on the heat flux incident on the unburnt material in the combusting zone due to flame from the pyrolysis zone. For small and laminar flames, spread rate follows x_p, $x_f \propto t^2$ while for turbulent and large flames spread rate is much higher and grows exponentially with time.

6.6 Dynamics of Cavity Fires

As discussed earlier, the air cavity is designed to facilitate proper ventilation and air circulation within the façade, as illustrated in Figs. 6.8 and 6.9. Nevertheless, in case of a fire, this space can act as a chimney and facilitate the upward spread of fire through a phenomenon known as the "chimney effect" [4, 19]. In many cases, direct exposure can also cause fire spread inside the cavity. Enclosure fires are often fuel

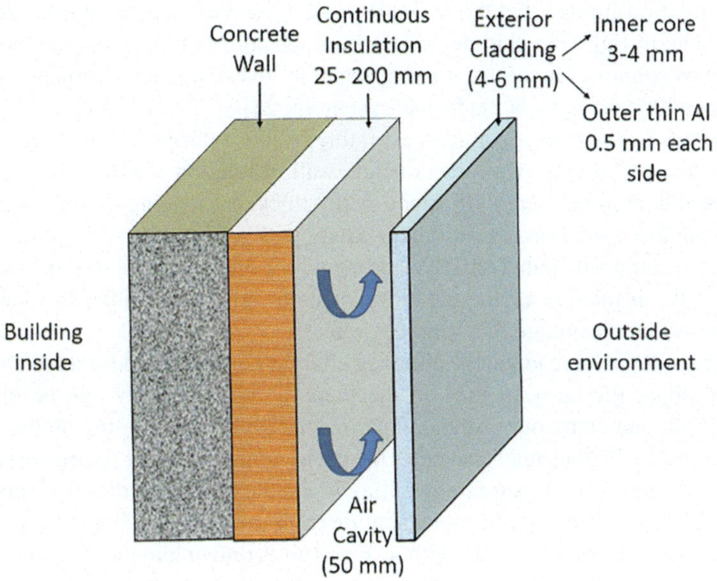

Fig. 6.8 Components of façade assembly [12, 30]

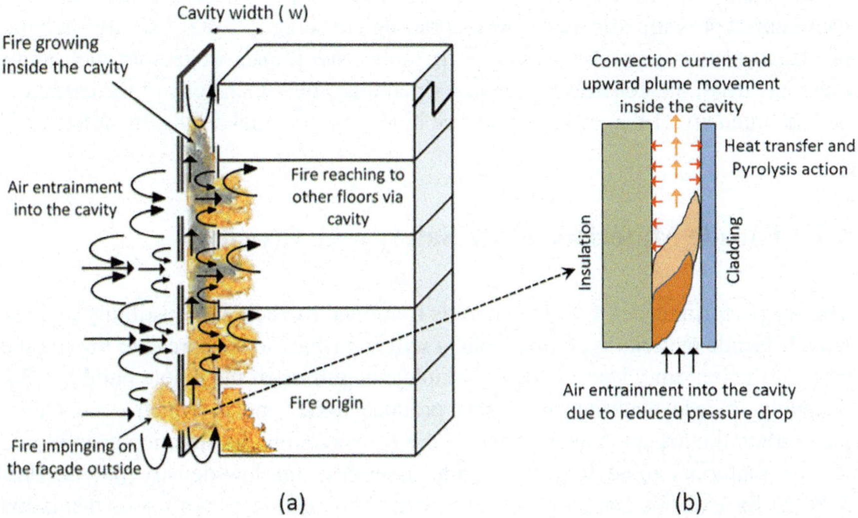

Fig. 6.9 (**a**) Chimney effect in façade fires, (**b**) Zoomed in view of processes taking place inside cavity [9, 30]

and ventilation-limited, but this is not the case for cavity fires, as ample ventilation is available through the chimney effect from the surroundings, making them challenging to contain. Therefore, it is important to investigate the chimney effect and corresponding fire risks in the façade system [5, 23].

Figure 6.8 illustrates a common assembly for the exterior wall or façade system of a high-rise building. After the concrete wall, insulation is added to enhance the R-value and improve energy efficiency, with a thickness ranging from 25 to 200 mm. Following the insulation layer is an air cavity with a thickness of 50 mm, and then aluminum composite panel (ACP) cladding with a thickness of 4–6 mm is attached. If either the insulation or the cladding, or both, are combustible, this can create conditions that encourage the "chimney effect".

In Fig. 6.9, fire spread via the chimney effect is demonstrated for a typical façade system where the buoyant pull on the flame along the cavity can be observed. Internal fire can come out of the compartment openings impinging on the exterior wall assembly. Flames may also enter the cavity region, heating the enclosed air far beyond the ambient temperature resulting in a reduction of air density. This causes initiation of upward buoyant movement or chimney effect leading to vacuum-like conditions inside the cavity of width w. Fresh air is drawn into the air gap created in between due to chimney effect and corresponding fire spread rate as:

$$Y = f\left(\Delta T, h, w, U\right)$$

This indicates that for a fixed sidewall height, the temperature difference can create the required pressure drop and flow rate through the cavity. To explore fires in building facades, it is crucial to examine the collective impact of factors such as the materials used, the construction design (including the dimensions of the chimney), and the intensity of the initial fire source in an experimental arrangement [9, 30].

6.7 Façade Materials – Fire Safety and Toxicity

The fire performance of façade materials is defined by their combustibility and toxicity. If façade material is combustible, it will contribute to the spread of fire to other parts of the building, leading to the complete engulfing of the whole building [34]. Secondly, façade materials may also produce toxic smoke, which causes more deaths than the fire itself as it spreads quickly, worsening the situation even more.

The materials most used in façade assembly are low-density polyethylene (LDPE) for exterior cladding and EPS, PIR, MW, etc., for continuous insulation. The problem with LDPE is that heat of formation of PE is very high while the heat of formation of CO_2 and H_2O is very low, so burning PE is a highly exothermic reaction (43.3 MJ/kg). Moreover, as mentioned in Table 6.1, it is noticeable that combustible insulation materials possess high R-values, which is why they are more

Table 6.1 Different continuous insulations used in façade assembly [30]

Continuous insulation	R-value (h.ft^2.^{0}F/BTU) per inch	Combustibility	Water permeability
Mineral wool	4.0	Low	High
PIR foam	6.5	Medium	Low
Phenolic foam	5.0–8.0	Medium	Low
EPS	5.0	High	Low

frequently utilized as continuous insulation in tall buildings. They can meet climate action SDGs of the Paris Agreement by making the building energy efficient [1]. Nevertheless, past incidents have demonstrated that such materials accelerate the spread of fire on building facades and generate poisonous fumes, as exemplified by the Grenfell Tower tragedy in 2017 [7, 35]. It greatly impacts the goal 11 of SDGs that focuses on making building safer [2].

6.8 Artificial Intelligence/Deep Learning Framework for Early Warning and Fire Risk Assessment

As described, façade fire behaviors are complex and could be affected by many factors, e.g., façade materials, ambient wind, façade type, and so on. Great efforts have been made to build mathematical and physical models for façade fire characteristics with these factors and deepen the understanding of the façade fire phenomenon. Nevertheless, in a real fire accident, many input parameters are unknown or unmeasurable to support the calculation of empirical equations. Also, the estimation depends highly on the user's knowledge and experience. A rapidly developed building fire proposes the requirement of a fire prediction tool which can achieve superfast response (at second level) with high accuracy based on limited data. Therefore, the artificial intelligence (AI)/deep learning (DL) tools are introduced in firefighting research to accomplish the task to identify and predict the building fire.

Figure 6.10 presents the framework to identify a compartment fire scenario and predict flashover occurrence with IoT sensors and a pre-trained DL model. The aim of detecting fires can be accomplished by using heat detectors, which gather temperature data via an IoT sensor network. The collected data is then transmitted to a cloud server [26, 27], where it can be managed, stored, and accessed by an AI engine that employs a Convolutional-LSTM network to identify the fire state and issue real-time alerts to occupants and firefighters regarding potential fire risks. Finally, a user interface (UI) is created to display the fire information, including measured data and AI outputs, to facilitate cyber-physical interaction.

Validated by a set of multi-scale fire tests, including a large-scale fire test in a 7.5 (L) × 3.4 (W) × 5.4 (H) m^3 chamber and a roughly 1/5 reduced-scaled model chamber, the proposed smart fire prediction system can identify the fire scenario

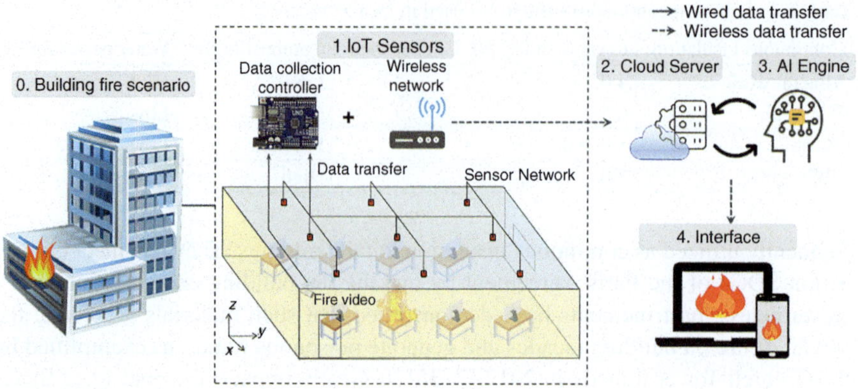

Fig. 6.10 Framework for IoT and DL model-based building fire identification and prediction [36]

(including the fire location and fire heat release rate) with an overall accuracy over 85% [36], and predict the onset of flashover with a lead time of 20 s [37]. This can provide important information to make firefighting decisions in the pre-flashover stage, and hence prevent the occurrence of façade fire.

For the post-flashover stage, the external fire (both the spilled flame and façade flame) is dominant. It is worth noting that the flame parameter(s) are difficult to measure by temperature or gas sensors while the fire images are commonly available. Thus, computer vision methods can be applied to analyze the façade fire features, as displayed in Fig. 6.11. A big database of 112 fire tests from the NIST Fire Calorimetry Database [38] is formed and 69,662 fire scene images labeled by their transient heat release rate (measured by the oxygen calorimetry) are adopted to train a CNN-based (Convolutional Neural Network) DL model.

The AI-image fire calorimetry approach is then employed to estimate the transient fire heat release rates (HRRs) for tests conducted in both the same and new laboratory environments, as well as real-world fire incidents. The outcomes demonstrate that the AI-image fire calorimetry technique can accurately predict the fire HRRs based on flame images, with a high degree of accuracy (coefficient of determination >0.8), irrespective of image background, camera settings, or viewing angles. This vision-based model exclusively uses fire images as inputs, offering an alternative means of determining the fire HRR through fire scene images when conventional calorimetric methods are impractical. For the post-flashover fire stage, the proposed model can be installed in a smartphone or carried by a UAV (Unmanned Aerial Vehicle) to get the façade fire picture and predict the fire heat release rate for fire risk assessment.

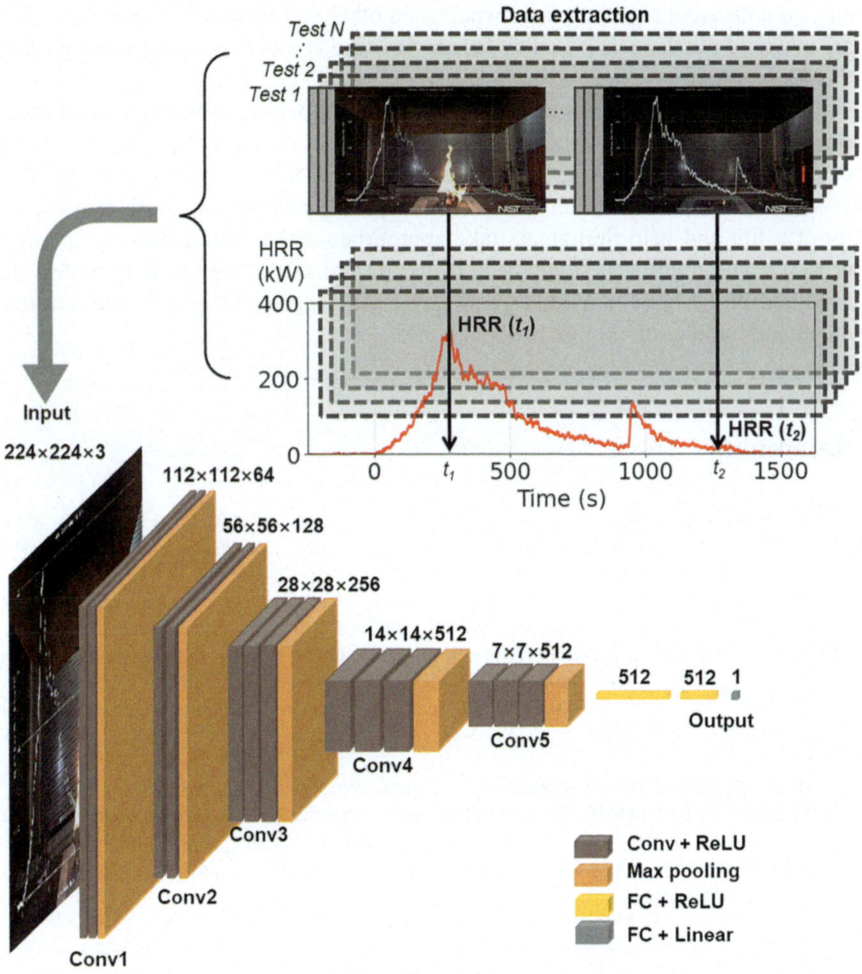

Fig. 6.11 Computer vision methods to analyze façade fire features [39]

6.9 Concluding Remarks

In light of the recent incidents involving façade fires and the importance of achieving the SDGs outlined in the Paris Agreement, it is imperative that we prioritize the study of various mechanisms that contribute to these types of fires. This will require diligent research and analysis to identify potential areas of improvement, as well as the development and implementation of robust safety standards and regulations to mitigate the risks associated with façade fires. This research can include studying the materials used in building façades, examining the design and construction of buildings, and investigating the causes of past façade fires. It is also important to engage with architects, building owners, and other stakeholders to ensure that they are aware of the dangers posed by façade fires and are taking appropriate measures to address them.

In addition to these measures, we can explore innovative technologies and materials that can help mitigate the risk of façade fires. For example, AI solution can help detect and respond to these fires more quickly and effectively. By analyzing data from sensors and cameras, an AI system can quickly identify the location and severity of a fire and help firefighters take appropriate action. Ultimately, by taking a proactive and multifaceted approach to façade fire safety, we can help protect the lives and property of individuals and communities, promote sustainable development, and build a safer, more resilient world.

References

1. The 17 GOALS | Sustainable Development. https://sdgs.un.org/goals. Accessed 11 Feb 2023.
2. Goal 11 | Department of Economic and Social Affairs. https://sdgs.un.org/goals/goal11. Accessed 11 Feb 2023.
3. Sun, J., Hu, L., & Zhang, Y. (2013). A review on research of fire dynamics in high-rise buildings. *Theoretical and Applied Mechanics Letters., 3*, 042001. https://doi.org/10.1063/2.1304201
4. Hu, L., Milke, J. A., & Merci, B. (2017). Special issue on fire safety of high-rise buildings. *Fire Technology, 53*, 1–3. https://doi.org/10.1007/S10694-016-0638-7/METRICS
5. Ronchi, E., & Nilsson, D. (2013). Fire evacuation in high-rise buildings: a review of human behaviour and modelling research. *Fire Science Reviews, 2*, 1–21. https://doi.org/10.1186/2193-0414-2-7
6. Ahrens, M. (2016). *High-rise building fires*. www.nfpa.org/osds. Accessed 6 May 2023.
7. McKenna, S. T., Jones, N., Peck, G., Dickens, K., Pawelec, W., Oradei, S., Harris, S., Stec, A. A., & Hull, T. R. (2019). Fire behaviour of modern façade materials – Understanding the Grenfell Tower fire. *Journal of Hazardous Materials, 368*, 115–123. https://doi.org/10.1016/j.jhazmat.2018.12.077
8. Jones, N., Peck, G., McKenna, S. T., Glockling, J. L. D., Harbottle, J., Stec, A. A., & Hull, T. R. (2021). Burning behaviour of rainscreen façades. *Journal of Hazardous Materials, 403*, 123894. https://doi.org/10.1016/j.jhazmat.2020.123894
9. Sharma, A. (2021). *Experimental and numerical investigations on external and internal fire spread in high-rise buildings*. PhD thesis, Indian Institute of Technology Roorkee. https://www.researchgate.net/publication/367389394_Experimental_and_numerical_investigations_on_external_and_internal_fire_spread_in_high-rise_buildings. Accessed 1 May 2023.
10. Gandhi, P., Jagdish, V., Karthikeyan, G., Chakravarthy, A., Nakrani, D., Ghoroi, C., & Srivastava, G. (2017). Performance of glass-ACP façade system in a full-scale real fire test in a G+2 structure. *Procedia Engineering, 210*, 512–519. https://doi.org/10.1016/J.PROENG.2017.11.108
11. Anderson, J., Boström, L., Jansson McNamee, R., & Milovanović, B. (2018). Experimental comparisons in façade fire testing considering SP Fire 105 and the BS 8414-1. *Fire and Materials, 42*, 484–492. https://doi.org/10.1002/FAM.2517
12. Bonner, M., & Rein, G. (2018). Flammability and multi-objective performance of building façades: Towards optimum design. *The International Journal of High-Rise Buildings, 7*, 363–374.
13. Sharma, A., & Mishra, K. B. (2023). A numerical case study on optimizing the smoke extraction fans in a model high-rise building fire. *Lecture Notes in Mechanical Engineering*, 405–417. https://doi.org/10.1007/978-981-19-6945-4_29/TABLES/1

14. Sharma, A., & Bhushan Mishra, K. (2019). Numerical studies on different stairwell smoke extraction techniques in a high-rise building. In *Proceedings of the Ninth International Seminar on Fire and Explosion Hazards (ISFEH9)* (Vol. 1, pp. 21–26). https://doi.org/10.18720/SPBPU/2/K19-114
15. Khan, A. A., Lin, S., Huang, X., & Usmani, A. (2023). Facade fire hazards of bench-scale aluminum composite panel with flame-retardant Core. *Fire Technology, 59*, 5–28. https://doi.org/10.1007/S10694-020-01089-4/FIGURES/9
16. Li, Y., Wang, Z., & Huang, X. (2022). An exploration of equivalent scenarios for building facade fire standard tests. *Journal of Building Engineering, 52*, 104399. https://doi.org/10.1016/J.JOBE.2022.104399
17. Anderson, J., Boström, L., Jansson McNamee, R., & Milovanović, B. (2018). Modeling of fire exposure in facade fire testing. *Fire and Materials, 42*, 475–483. https://doi.org/10.1002/FAM.2485
18. Guillaume, E., Fateh, T., Schillinger, R., Chiva, R., & Ukleja, S. (2018). Study of fire behaviour of facade mock-ups equipped with aluminium composite material-based claddings, using intermediate-scale test method. *Fire and Materials, 42*, 561–577. https://doi.org/10.1002/FAM.2635
19. Lugaresi, F., Kotsovinos, P., Lenk, P., & Rein, G. (2022). Review of the mechanical failure of non-combustible facade systems in fire. *Construction and Building Materials, 361*, 129506. https://doi.org/10.1016/J.CONBUILDMAT.2022.129506
20. Babrauskas, V. (1996). Façade fire tests: Towards an international test standard. *Fire Technology, 32*, 219–230. https://doi.org/10.1007/BF01040215/METRICS
21. Barbara Lane. (2018). Dr Barbara Lane's expert report | Grenfell Tower Inquiry, Grenfell Tower Inquiry.
22. Drysdale, D. (2011). *An introduction to fire dynamics* (3rd ed.). Wiley.
23. Hurley, M. J., Gottuk, D. T., Hall, J. R., Jr., Harada, K., Kuligowski, E. D., Puchovsky, M., Watts, J. M., Jr., & Wieczorek, C. J. (2015). *SFPE handbook of fire protection engineering* (5th ed.). Springer.
24. Sharma, A., & Bhushan Mishra, K. (2019). Experimental set-up to measure the maximum mass burning rate of storage tank fires. *Process Safety and Environmental Protection., 131*, 282–291. https://doi.org/10.1016/j.psep.2019.09.001
25. Sharma, A., Vishwakarma, P. K., Mishra, S., & Mishra, K. B. (2020). Visible length and thermal radiation of WCO biodiesel and ethanol pool fires. *Fuel, 280*, 118396. https://doi.org/10.1016/j.fuel.2020.118396
26. Mishra, S., Vishwakarma, P. K., Sharma, A., & Mishra, K. B. (2021). Experimental investigation of small-scale CS2 (carbon disulphide) pool fires. *Process Safety and Environmental Protection, 145*, 203–210. https://doi.org/10.1016/J.PSEP.2020.08.002
27. Kawagoe, K. (1958). Fire behavior in rooms. *Report 27, Building Research Institute,* Ministry of Construction, Tokyo, Japan.
28. Mizuno, T., & Kawagoe, K. (1986). Burning behaviour of upholstered chairs. Part 3. Flame and plume characteristics in fire test. *Fire Science and Technology, 6*, 29–37. https://doi.org/10.3210/fst.6.29
29. Kotthoff, I., & Riemesch-Speer, J. (2013). Mechanism of fire spread on facades and the new Technical Report of EOTA "Large-scale fire performance testing of external wall cladding systems". In *MATEC web of conferences, EDP sciences* (p. 2010).
30. Sharma, A., & Mishra, K. B. (2021). Experimental investigations on the influence of 'chimney-effect' on fire response of rainscreen façades in high-rise buildings. *Journal of Building Engineering, 44*, 103257. https://doi.org/10.1016/J.JOBE.2021.103257
31. Orloff, L., Modak, A. T., & Alpert, R. L. (1977). Burning of large-scale vertical surfaces. *Symposium (International) on Combustion, 16*, 1345–1354. https://doi.org/10.1016/S0082-0784(77)80420-7
32. Alpert, R. L., & Ward, E. J. (1984). Evaluation of unsprinklered fire hazards. *Fire Safety Journal, 7*, 127–143. https://doi.org/10.1016/0379-7112(84)90033-X

33. Gollner, M. J., Overholt, K., Williams, F. A., Rangwala, A. S., & Perricone, J. (2011). Warehouse commodity classification from fundamental principles. Part I: Commodity & burning rates. *Fire Safety Journal, 46*, 305–316.
34. Srivastava, G., & Gandhi, P. D. (2022). *Performance of combustible façade systems used in green building technologies under fire*. https://doi.org/10.1007/978-981-16-3112-2.
35. Khan, A. A., Lin, S., Huang, X., & Usmani, A. (2021). Facade fire hazards of bench-scale aluminum composite panel with flame-retardant core. *Fire Technology*, 1–24. https://doi.org/10.1007/s10694-020-01089-4
36. Zhang, T., Wang, Z., Zeng, Y., Wu, X., Huang, X., & Xiao, F. (2022). Building artificial-intelligence digital fire (AID-fire) system: A real-scale demonstration. *Journal of Building Engineering, 62*, 105363. https://doi.org/10.1016/J.JOBE.2022.105363
37. Zhang, T., Wang, Z., Wong, H. Y., Tam, W. C., Huang, X., & Xiao, F. (2022). Real-time forecast of compartment fire and flashover based on deep learning. *Fire Safety Journal, 130*, 103579. https://doi.org/10.1016/J.FIRESAF.2022.103579
38. Fire Calorimetry Database (FCD) | NIST. https://www.nist.gov/el/fcd. Accessed 1 May 2023.
39. Wang, Z., Zhang, T., Wu, X., & Huang, X. (2022). Predicting transient building fire based on external smoke images and deep learning. *Journal of Building Engineering, 47*, 103823. https://doi.org/10.1016/J.JOBE.2021.103823

Chapter 7
Experimental Study on Self-Healing of Micro-Cracks in Concrete with Combination of Environmentally Friendly Bacteria

R. Anjali, S. Anandha Kumar, Jaswanth Gangolu, and R. Abiraami

7.1 Introduction

Concrete is the most extensively used material in building construction globally. However, concrete is weak in tension, ductility, and crack resistance. In addition, cracks can form in concrete structures due to external loads, drying shrinkage, and freeze–thaw action [1, 2]. These cracks provide a pathway for the deterioration of the structures by generating external moisture and corrosion in the reinforcement materials which reduces the life of concrete [3, 4]. Hence, it is important to repair cracks in the structures effectively. The self-healing concrete (SHC) mechanism resembles how the human body naturally repairs wounds over time. The ability of concrete to self-heal cracks would increase the structure's durability and sustainability, extending its service life. Study on autonomous SHC was done in 1994 [5, 6]. SHC is a concrete composite that may automatically cure minor cracks without the need for external diagnosis or human intervention [7]. Cracks in concrete can be filled by materials such as cement or resins. Cement grout is an effective method to fill cracks but it is unable to penetrate fine cracks which are less than a millimeter wide [8]. Therefore, various studies have been carried out globally to overcome this problem leading to the discovery that microbial mineral precipitation acts to heal concrete cracks [9]. Biocementation (bacteria-based method) has been implemented to heal concrete cracks. Biocementation is an environmental friendly and economic process in which bacteria produce the calcium carbonate precipitate at the cracked

R. Anjali
School of Civil Engineering, Vellore Institute of Technology, Vellore, Tamil Nadu, India

S. Anandha Kumar (✉) · J. Gangolu
Department of Civil Engineering, Aditya Engineering College, Surampalem, India

R. Abiraami
Department of Civil Engineering, Christian College of Engineering and Technology, Oddanchatram, Dindigul, Tamil Nadu, India

A. Bahrami (ed.), *Sustainable Structures and Buildings*,
https://doi.org/10.1007/978-3-031-46688-5_7

zone [8]. The carbonate precipitate is used in many areas such as the treatment of soil, heavy metal remediation, cementing of sandy soil, etc [10, 11]. Recently, calcium precipitate has been used in many engineering applications and is referred to as the microbial-induced calcite precipitation (MICP) technique. This method is used in concrete cracks to improve concrete durability [12–14].

Compared to conventional concrete, sustainable concrete uses much less energy during manufacture and generates significantly less carbon dioxide. Currently, Portland cement is an extensively utilized material in the manufacture of concrete. Some geographic locations are quickly running out of limestone resources for cement production. Large urban regions are running out of materials to use as aggregates in concrete production. Sustainability necessitates that those in the construction sector consider the whole life cycle of a structure, including construction, maintenance, demolition, and recycling [15–17].

Different types of bacteria (*Bacillus subtilis, Bacillus cereus, Bacillus sphaericus,* etc.) are used for mineral precipitation. This *Bacillus* group has the ability to function in a concrete environment due to the high alkalinity of concrete and the alkaliphilic strains [8]. Through the synthesis of the urease enzyme, ureolytic bacteria precipitate calcium carbonate. The hydrolysis of urea into CO_2 and ammonia, which is catalyzed by this enzyme, raises the pH and increases the concentration of carbonate in the bacterial environment. Sand columns have been used to precipitate calcium carbonate using biodeposition technology [18, 19] and for remediation of cracks in concrete [20]. The MICP technique is a new approach to the remediation of cracks and to protect the serviceability of concrete [1]. *Bacillus subtilis* has been mixed in concrete and was studied for the compressive strength and self-healing properties of concrete. The results revealed that the compressive strength of concrete was noticeably increased and attained the self-healing property due to the continuous precipitation of calcite [21]. Calcite ($CaCO_3$) may be determined by four important variables, including pH, calcium concentration [22], dissolved inorganic carbon concentration, and the existence of nucleation sites [22]. Salmasi and Mostofinejad [23] studied the effects of bacteria on enhancing the strength of cement and mortar using varied concentrations of anaerobic bacteria (*Shewanella* species) and observed strength improvements in their specimens.

A study was conducted to find the effects of different microbes, properties, and efficiency, in addition to an overview of microbial concrete uses [11]. Two types of bacteria (*Sporosarcina pasteurii* and *Bacillus subtilis*) were employed in the study to observe the influence of bacteria on concrete permeability, corrosion, and electrical resistivity in two different mix designs. The results showed decreased water absorption and chloride penetrability and that *Sporosarcina pasteurii* enhances the compressive strength and electrical resistivity in concrete [23]. Researchers concentrated on different ureolyic bacteria similar to *Sporosarcina pasteurii, Bacillus sphaericus,* and *Bacillus megaterium*, which are alkali-resistant and have the spore-forming capability to help them live up without nutrients [24–26]. The researchers investigated the impact of calcium acetate, calcium nitrate, calcium chloride, and calcium oxide on MICP by *Bacillus alkalinitrilicus* and *Bacillus subtilis* [27–29]. With and without the inclusion of external calcium sources, the optimum amount to improve the compressive strength was examined utilizing *Bacillus subtilis* [30]. The earlier research clearly demonstrates that bacteria utilize the ureolysis, denitrification, and metabolic conversion of the organic chemical used in microbial concrete as per the MICP processes. The current study proposed to investigate the effect of bacteria (*Bacillus subtilis* and *Bacillus cereus*) on its properties, and self-healing of cracks in concrete compared to conventional concrete.

In the subsequent years, several researchers began to work on this topic. Research on SHC was begun in 2006 [31, 32]. After conducting experimental investigations, the *Bacillus* genus was selected as an ideal SHC therapeutic agent [32].

7.2 Materials

7.2.1 Cement, Aggregates, and Water

Ordinary Portland cement 43 grade was utilized in the study as per Indian standard specifications (IS 8112:1989). The physical properties of cement are presented in Table 7.1. The fine aggregates up to a maximum size of 4.75 mm and coarse aggregates passed through 20 mm and retained on 12.5 mm were used in this study. The specific gravity values of fine and coarse aggregates were 2.66 and 2.74, respectively. Distilled water was utilized for mixing and curing of specimens as well as the preparation of nutrient broth solutions for bacterial species, and casting and curing of microbial samples. The standard M25 grade concrete mixes were prepared.

7.2.2 Selection of Bacteria

By using biomineralization, organisms like *Bacillus* enabled the formation of microbial concrete. In this study, two species were chosen, *Bacillus subtilis* and *Bacillus cereus* obtained from Manidharma Biotech Private Limited, Chennai. These species were gram-positive rod-shape structures and tolerate extreme environmental conditions. It could survive in high alkali conditions in concrete since the formation of $CaCO_3$ signified the presence of bacteria. The culture was kept alive on nutrient agar slants and subcultured on the sterilized medium every 20 days.

Table 7.1 Physical properties of cement

Properties	Values
1. Specific gravity	3.15
2. Consistency (%)	33
3. Initial setting time (min)	87
4. Final setting time (min)	575

7.2.3 Preparation of Bacterial Solution

Bacillus subtilis and *Bacillus cereus* were developed in the Lysogeny broth (LB) medium. LB was nutrition-rich and was primarily used for the growth of bacteria. These species were inoculated in the LB medium for multiplying the growth of bacteria. Then, 12.5 g of LB medium was added to a 250 ml conical flask containing distilled water. LB medium consisted of 10 g of tryptone, 5 g of yeast extract, and 10 g of sodium chloride in 1 L medium. Thereafter, the conical flask was cotton plugged, covered with paper, and a rubber band made this airtight. The solution was sterilized in an autoclave for 10–20 min approximately. The sterilization temperature was maintained at 120 °C for 20 min. The solution was free from contaminants and pure orange in color. The bacteria were added to the media with the help of an inoculation loop. The solution was kept undisturbed for 24 h in the laminar flow chamber. The bacterial solution contained turbidity that indicated the growth of bacteria which was different from the control. After 20 days, it was subcultured in the sterilized medium. The preparation of bacterial solution is displayed in Fig. 7.1.

7.2.4 Preparation of Bacterial Concrete

7.2.4.1 Mix Design

Table 7.2 lists the material requirements for both conventional and bacterial concrete. Four concrete mixtures were chosen for research on the self-healing characteristics of concrete. The proportions of cement, sand, coarse aggregate, and water used in each example were identical. Blend 1 was a representation of nominal concrete with no bacterial broth solution added. In the case of Blend 2, 250 ml of *Bacillus cereus* was added to Blend 1. Similarly, 250 ml of *Bacillus subtilis* was added to Blend 1 to form Blend 3. Finally, 250 ml of a bacterial solution containing *Bacillus subtilis* and *Bacillus cereus* was added to Blend 1, resulting in Blend 4.

Fig. 7.1 Preparation of bacterial solution: (**a**) Subcultured *Bacillus subtilis* bacterial solution, (**b**) Subcultured *Bacillus cereus* bacterial solution

Table 7.2 Mix design of nominal and bacterial concrete

Materials	Conventional concrete (Blend 1)	*Bacillus cereus* with conventional concrete (Blend 2)	*Bacillus subtilis* with conventional concrete (Blend 3)	Combined bacterial with conventional concrete (Blend 4)
Cement (kg)	345	345	345	345
Sand (kg)	750	750	750	750
Coarse aggregate (kg)	1158	1158	1158	1158
Water (m^3)	0.186	0.186	0.186	0.186
Bacterial broth added	Nil	*Bacillus cereus*	*Bacillus subtilis*	*Bacillus subtilis* + *Bacillus cereus*
Bacterial broth solution (ml)	Nil	250	250	250

7.2.4.2 Mixing

Mixing of concrete was done employing a motorized electrical mixer. The required quantities of cement, fine aggregate, and coarse aggregate were weighed and put in uniform layers. The dry mixing was done to get a homogeneous mixture. Then, the estimated quantity of bacterial solution along with water was added and mixed for up to 5 min to obtain the homogeneity. This fresh concrete was used to verify the workability of concrete immediately before casting.

7.2.4.3 Casting

Iron molds of different sizes were used to cast the concrete specimens for the compressive strength test, splitting tensile strength test, and flexural strength test. Initially, the mold was checked for joint movement and lubricated with oil. The prepared fresh concrete was poured into the cube, cylinder, and prism of size 150 mm × 150 mm × 150 mm, 150 mm × 300 mm, and 100 mm × 100 mm × 500 mm, respectively, and compacted manually in three layers. The prepared specimens are depicted in Fig. 7.2.

7.3 Experimental Investigation

In order to determine the slump value, a number of experiments were conducted (ASTM-C143, 2000) on the compressive strength (ASTM-C873, 2000 and ASTM-C943, 2000), splitting tensile strength (ASTM-C496/C496M, 2000), and flexural strength (ASTM-D790, 2000) of different mixes (i.e., conventional and

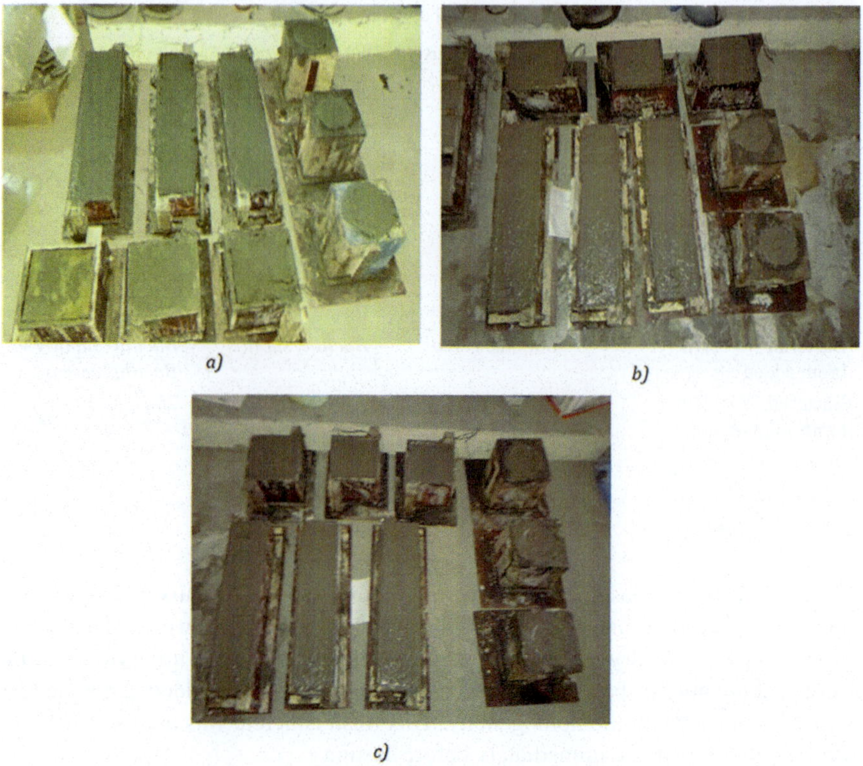

Fig. 7.2 Casting of concrete specimens: (a) *Bacillus subtilis* concrete specimens, (b) *Bacillus cereus* concrete specimens, (c) *Bacillus subtilis* and *Bacillus cereus* concrete specimens

bacteria concrete) with various curing periods of 7, 14, and 28 days [33–37]. Figure 7.3 illustrates the experimental test setups. Three samples for each test were evaluated to ensure the accuracy and repeatability of the results.

7.4 Results and Discussion

7.4.1 Slump Value

The slump cone test was utilized to determine the workability of the concrete mix. True slump, as opposed to shear and collapsible slump, was the desired shape of the slump. The lower the value of the slump, the more workable the mix. Low-slump concrete is stiff, dry, and difficult to work with. Likewise, if there is a significant slump, it can be due to high water content which eventually affects the strength and workability of concrete. According to IS 456:2000, the slump ranges between 50 mm and 100 mm for ordinary reinforced concrete beams and slabs. However, the

Fig. 7.3 Experimental test setups: (**a**) Compressive strength test, (**b**) Splitting tensile strength test, (**c**) Flexural strength test

Table 7.3 Slump value of different concrete mixes

Concrete mix	Slump value (mm)
Blend 1	90
Blend 2	85
Blend 3	80
Blend 4	70

selected four concrete mixes, Blends 1, 2, 3, and 4, the slump values of 90 mm, 85 mm, 80 mm, and 70 mm, respectively (Table 7.3). According to ASTM-C143, the slump value of concrete used in the original works of beams and slabs is 50–100 mm. Figure 7.4 shows measuring of the slump value.

A 250 ml addition of *Bacillus subtilis* to the nominal concrete mix reduced the slump by 11.11%. However, adding 250 ml of *Bacillus cereus* to the nominal mix reduced the slump value by 5.55%. *Bacillus subtilis* bacterial broth produced a better concrete slump than *Bacillus cereus*. Furthermore, adding 250 ml of bacterial broth, *Bacillus subtilis* and *Bacillus cereus*, to the nominal mix substantially reduced the slump value by 22.22%. Based on the results, the combined action of the selected bacteria created an efficient slump value that can be employed quickly and economically. In the instance of Blend 4, the consistency of fresh concrete, before it set,

Fig. 7.4 Measuring true slump value

was good. Furthermore, the amount of water required was smaller than in the other three scenarios, which could result in a sustainable and cost-effective outcome. The lower the slump, the harder and less workable the concrete would be, but the hardness should not be much less as it would be difficult to construct. Blend 4 had greater strength and durability than other cases, as it is commonly known that the higher the slump, the lower the strength and durability.

7.4.2 Compressive Strength

The compressive strength of concrete is an important parameter to evaluate the concrete performance. The individual and combined effects of both bacteria on the compressive strength of conventional concrete are demonstrated in Fig. 7.5. It is observed that the effect of *Bacillus subtilis* exhibited better results than *Bacillus cereus*, and increased the compressive strength of conventional concrete considerably from 25 MPa to 33 MPa (i.e., 29.8%), whereas *Bacillus cereus* provided a 4% increase in the compressive strength after 28 days of curing.

The combined effect of both bacteria on conventional concrete gave a better compressive strength of 29.7 MPa (i.e., 17.9%) when compared to *Bacillus cereus* blended concrete after 28 days of curing. This obtained value of the compressive strength was slightly lower (i.e., 9.17% decrease) than the compressive strength of Blend 3. From this study, it is evident that *Bacillus subtilis* alone produced better strength than the combination of *Bacillus subtilis* and *Bacillus cereus*. *Bacillus subtilis* remarkably increased the strength of concrete and hastened the healing of micro-cracks in the samples. Figure 7.5 indicates that the strength at 7 days for Blend 3 specimen and other cases varied noticeably, and a comparable difference

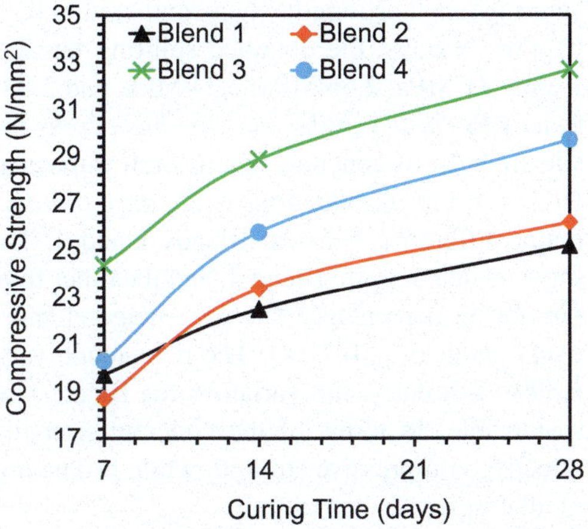

Fig. 7.5 Results of compressive strength test

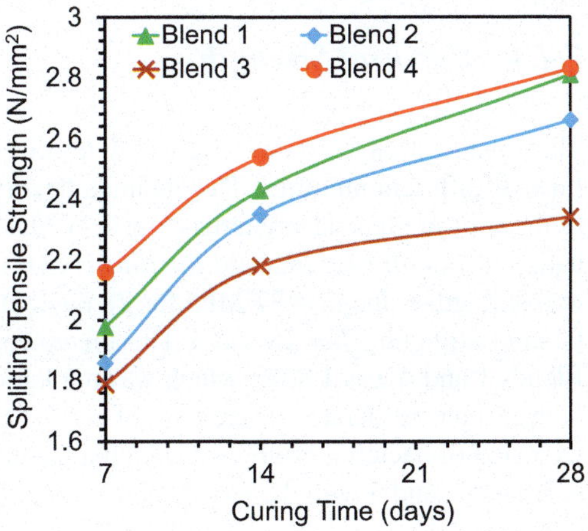

Fig. 7.6 Results of splitting tensile strength test

was witnessed at 28 days as well. The compressive strength is an important factor for bearing greater loads over time; by adding 250 ml *Bacillus subtilis* solution to the standard concrete specimen, the structure's sustainability is improved. In this sense, the use of concrete and the inclusion of reinforcement can be greatly reduced while maintaining the same strength and durability.

7.4.3 Splitting Tensile Strength

Regardless of grade, the concrete's tensile behavior is always poor. However, concrete has some tensile strength that is only substantial when compared to the compressive strength. The current study tested the tensile behavior of concrete for four chosen blend types (Fig. 7.6). The nominal concrete mix (Blend 1) exhibited better splitting tensile strength when compared to individual bacterial broth (Blends 2 and 3). For 28 days, the obtained splitting tensile strengths of Blends 1, 2, 3, and 4 were 2.81 MPa, 2.66 MPa, 2.34 MPa, and 2.83 MPa, respectively. In fact, the addition of *Bacillus subtilis* and *Bacillus cereus* individually reduced the splitting tensile strength of concrete. The overall impact was greater splitting tensile strength. However, the introduction of *Bacillus subtilis* and *Bacillus cereus* eliminated the major difference between Blends 1 and 4. The gap between the maximum splitting tensile strength, Blend 4, and the other three cases appeared to be less by value but not by percentage. The percentage change between Blends 4 and 1 was relatively small (i.e., 0.71%). The percentage variation for Blend 4 and Blend 2 was 6.39%. Similarly, the variation for Blend 3 was significant at 20.94%, which is sustainable. In terms of the concrete strength, *Bacillus subtilis* developed more effective compressive strength while producing lower splitting tensile strength out of all cases.

7.4.4 Flexural Strength

The flexural strength of concrete was determined by doing the flexural strength tests for four different mixtures. The obtained flexural strengths followed the same trend as the splitting tensile strengths (Fig. 7.7). Out of four blends, the bacterial combination of *Bacillus subtilis* and *Bacillus cereus* (Blend 4) demonstrated a good flexural strength value of 4.37 MPa for 28 days. It is seen that Blend 1 had 4.29 MPa flexural strength after 28 days. The increment of the flexural strengths between Blends 1 and 4 was 1.86%, which was not large. However, the flexural strength for 7 days had moderate differences of 3.26 MPa and 3.53 MPa. The individual inclusion of bacterial broth worsened the flexural nature of concrete, as it is evident in Blends 2 and 3 with the strengths of 4.17 MPa and 3.98 MPa.

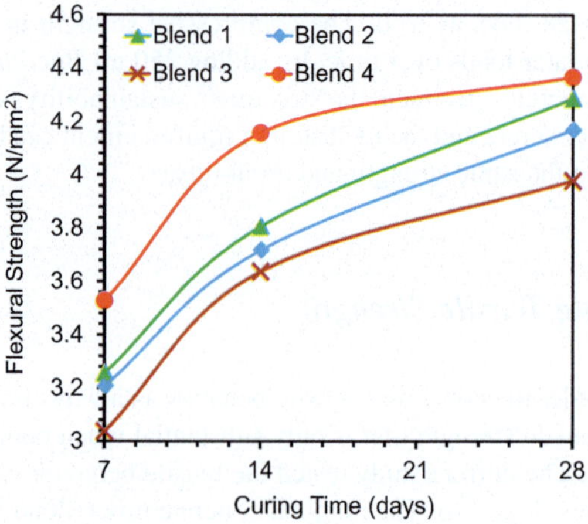

Fig. 7.7 Results of flexural strength test

7.4.5 *Self-Healing Characteristics of Combined Bacterial Concrete*

Concrete that is capable of self-healing or self-repairing cracks is known as self-healing or self-repair concrete. It not only closes cracks but also partially or completely restores the structure's mechanical properties. Concrete frequently develops surface cracks due to its low tensile strength in comparison to other construction materials. Because they allow the movement of liquids and gases that may contain toxic compounds, these cracks degrade the durability of concrete. If microcracks spread to steel reinforcements, not only concrete but also reinforcements will be vulnerable.

However, SHC is still in its early stages of development, with the current study focusing on healing of tensile cracks in concrete. In this study, the cracks formed in Blend 4, after the compressive strength testing, were observed for its self-healing nature by curing the cracked surface for 7 days. It was seen that the combination of two bacteria, *Bacillus subtilis* and *Bacillus cereus*, acted as a self-healing agent to arrest micro-cracks. When water was provided to bacterial concrete, it infiltrated through the cracks, creating the conditions for bacteria present in concrete to precipitate calcite. Calcite precipitation resulted in filling concrete cracks. Figure 7.8 displays the effect of calcite precipitation filling before and after healing. The figure illustrates a microcrack after performing a compressive strength test on Blend 4. This microcrack signified the failure of the concrete block after it reached its maximum strength. The addition of *Bacillus subtilis* and *Bacillus cereus* automatically cured the microcracks. This study revealed that not only were microcracks healed, but the slump value, splitting tensile strength, and flexural strength of concrete also improved. Thermal stresses and unexpected loads are known to cause the formation of microcracks in structures. However, for such cases, the Blend 4 scenario was the best and most sustainable combination.

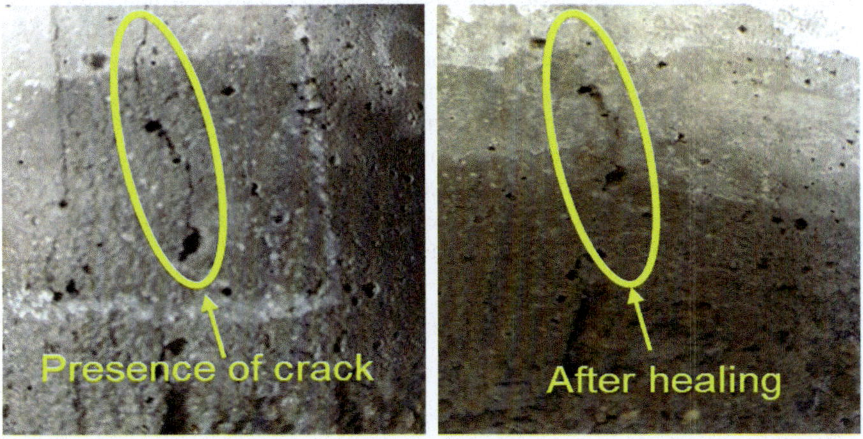

Fig. 7.8 Combined bacterial concrete before and after healing

7.5 Conclusions

The objective of this study was to examine healing of concrete microcracks with the addition of bacterial broth to the standard mix concrete. Four concrete mixes, i.e., nominal concrete, *Bacillus cereus* with concrete, *Bacillus subtilis* with concrete, and combined bacteria with concrete were used. The total amount of bacteria added to the nominal mix from Blends 2–4 was 250 ml. The slump cone test, compressive strength test, splitting tensile test, and flexural strength test were performed to determine the mechanical properties of the mixes. The slump values of 90 mm, 85 mm, 80 mm, and 70 mm were obtained. A good slump always results in a lower value, indicating the workability of concrete. However, out of all types, Blend 4 provided an efficient slump value with a 22.22% variance from the nominal mix. In the compressive strength test, out of all selected blends, Blend 3 produced substantial compressive strength with a 29.7% increase above the normal mix. Remarkably, Blend 4 yielded lower compressive strength than Blend 3. Blend 4 exhibited an excellent tensile nature in the splitting tensile test, however, the difference between Blends 4 and 1 was minor. The combined effect of bacteria had little effect on the splitting tensile strength of the mixture. Similar trend was resulted from the flexural strength test. Blend 4 coupled with *Bacillus subtilis* and *Bacillus cereus* healed the microcracks in concrete. When water was added to bacterial concrete, it infiltrated through the crevices, allowing bacteria in concrete to precipitate calcite. Calcite precipitation caused concrete cracks to fill. The chosen SHC blends healed and reduced the need for external intervention to locate and fix internal damage (e.g., cracks), which is a major benefit to the construction industry. In addition, these blends reduce reinforcement corrosion and concrete deterioration while decreasing costs and enhancing the durability. It has the potential to survive for decades or centuries. This is the most significant advantage of SHC. This guarantees that you will not have to replace the concrete surface during its lifetime. SHC requires less upkeep. Cracks in conventional concrete must be filled and sealed. SHC also improved the compressive strength of concrete which is a great advantage to industries while constructing. SHC has the potential to contribute to the infrastructure crisis. This approach can lower the environmental costs of frequent concrete upkeep, demonstrating its usefulness as a sustainable resource. This has social, economic, and environmental benefits, because eliminating or lowering the need for maintenance and/or improving longevity minimizes disruption, as well as costs and material use. With a similar process obtained in this study, the current work can be extended to assess the mechanical properties of concrete for bigger units such as beams, columns, and slabs subjected to a variety of loads.

References

1. Wu, M., Hu, X., Zhang, Q., Cheng, W., Xue, D., & Zhao, Y. (2020). *Application of bacterial spores coated by a green inorganic cementitious material for the self-healing of concrete cracks* (Vol. 113). Elsevier.
2. Madhan Kumar, M., Vijaya Ganapathy, D., Subathra Devi, V., & Iswarya, N. (2019). Experimental investigation on fibre reinforced bacterial concrete. *Materials Today Proceedings, 22*, 2779–2790. https://doi.org/10.1016/j.matpr.2020.03.409
3. Bhaskar, S., Anwar Hossain, K. M., Lachemi, M., Wolfaardt, G., & Otini Kroukamp, M. (2017). Effect of self-healing on strength and durability of zeolite-immobilized bacterial cementitious mortar composites. *Cement and Concrete Composites, 82*, 23–33. https://doi.org/10.1016/J.CEMCONCOMP.2017.05.013
4. De Muynck, W., Cox, K., De Belie, N., & Verstraete, W. (2008). Bacterial carbonate precipitation as an alternative surface treatment for concrete. *Construction and Building Materials, 22*(5), 875–885. https://doi.org/10.1016/j.conbuildmat.2006.12.011
5. Shivanshi, S., Chakraborti, G., Upadhyaya, K. S., & Kannan, N. (2023). A study on bacterial self-healing concrete encapsulated in lightweight expanded clay aggregates. *Materials Today: Proceedings. (In Press).* https://doi.org/10.1016/j.matpr.2023.03.541
6. Van Tittelboom, K., & De Belie, N. (2013). Self-healing in cementitious materials—A review. *Materials, 6*(6), 2182–2217.
7. Aytekin, B., Mardani, A., & Yazıcı, Ş. (2023). State-of-art review of bacteria-based self-healing concrete: Biomineralization process, crack healing, and mechanical properties. *Construction and Building Materials, 378*, 131198.
8. Kaur, N. P., Majhi, S., Dhami, N. K., & Mukherjee, A. (2020). Healing fine cracks in concrete with bacterial cement for an advanced non-destructive monitoring. *Construction and Building Materials, 242*, 118151. https://doi.org/10.1016/j.conbuildmat.2020.118151
9. Pachaivannan, P., Hariharasudhan, C., Mohanasundram, M., & Anitha Bhavani, M. (2020). Experimental anaylsis of self healing properties of bacterial concrete. *Materials Today Proceedings, 33*, 3148–3154. https://doi.org/10.1016/j.matpr.2020.03.782
10. Whiffin, V. S., van Paassen, L. A., & Harkes, M. P. (2007). Microbial carbonate precipitation as a soil improvement technique. *Geomicrobiology Journal, 24*(5), 417–423. https://doi.org/10.1080/01490450701436505
11. Chiung-Wen, C., Aydilek, A. H., Seagren, E. A., & Lai, M. (2011). Biocalcification of Sand through Ureolysis. *Journal of Geotechnical and Geoenvironmental Engineering, 137*(12), 1179–1189. https://doi.org/10.1061/(ASCE)GT.1943-5606.0000532
12. Wang, J., Ersan, Y. C., Boon, N., & De Belie, N. (2016). Application of microorganisms in concrete: A promising sustainable strategy to improve concrete durability. *Applied Microbiology and Biotechnology, 100*(7), 2993–3007. https://doi.org/10.1007/s00253-016-7370-6
13. Bang, S. S., Lippert, J. J., Yerra, U., Mulukutla, S., & Ramakrishnan, V. (2010). Microbial calcite, a bio-based smart nanomaterial in concrete remediation. *International Journal of Smart and Nano Materials, 1*(1), 28–39. https://doi.org/10.1080/19475411003593451
14. Achal, V., & Mukherjee, A. (2015). A review of microbial precipitation for sustainable construction. *Construction and Building Materials, 93*, 1224–1235. https://doi.org/10.1016/j.conbuildmat.2015.04.051
15. Naik, T. R. (2008). Sustainability of concrete construction. *Practice Periodical on Structural Design and Construction, 13*(2), 98–103.
16. McDonough, W. P. (1992). *The hannover principles: Design for sustainability. EXPO 2000, Hannover, Germany*. William Mcdonough & Partners.
17. Worrell, E., & Galtisky, C. (2004). *Energy efficiency improvement and cost saving opportunities for cement making*. Lawrence Berkeley National Laboratory Publication LBNL-54036.
18. Nemati, M., & Voordouw, G. (2003). Modification of porous media permeability, using calcium carbonate produced enzymatically in situ. *Enzyme and Microbial Technology, 33*(5), 635–642. https://doi.org/10.1016/S0141-0229(03)00191-1

19. Stocks-Fischer, S., Galinat, J. K., & Bang, S. S. (1999). Microbiological precipitation of CaCO3. *Soil Biology and Biochemistry, 31*(11), 1563–1571. https://doi.org/10.1016/S0038-0717(99)00082-6
20. Bang, S. S., Galinat, J. K., & Ramakrishnan, V. (2001). Calcite precipitation induced by polyurethane-immobilized *Bacillus pasteurii. Enzyme and Microbial Technology, 28*(4), 404–409. https://doi.org/10.1016/S0141-0229(00)00348-3
21. Kumar, A. P., Devi, A. R. A., Anestraj, S., Arun, S., & Santhoshkumar, A. (2015). An experimental work on concrete by adding *Bacillus subtilis. Indian Journal of Environmental Protection, 35*, 911–915.
22. Archana, G., Dhodapkar, R., & Kumar, A. (2016). Offline solid-phase extraction for preconcentration of pharmaceuticals and personal care products in environmental water and their simultaneous determination using the reversed phase high-performance liquid chromatography method. *Environmental Monitoring and Assessment, 188*(9). https://doi.org/10.1007/s10661-016-5510-1
23. Salmasi, F., & Mostofinejad, D. (2020). Investigating the effects of bacterial activity on compressive strength and durability of natural lightweight aggregate concrete reinforced with steel fibers. *Construction and Building Materials, 251*, 119032. https://doi.org/10.1016/j.conbuildmat.2020.119032
24. Wang, J. Y., Soens, H., Verstraete, W., & De Belie, N. (2014). Self-healing concrete by use of microencapsulated bacterial spores. *Cement and Concrete Research, 56*, 139–152. https://doi.org/10.1016/j.cemconres.2013.11.009
25. Van Tittelboom, K., De Belie, N., De Muynck, W., & Verstraete, W. (2010). Use of bacteria to repair cracks in concrete. *Cement and Concrete Research, 40*(1), 157–166. https://doi.org/10.1016/j.cemconres.2009.08.025
26. Andalib, R., et al. (2016). Optimum concentration of *Bacillus megaterium* for strengthening structural concrete. *Construction and Building Materials, 118*, 180–193. https://doi.org/10.1016/j.conbuildmat.2016.04.142
27. Khaliq, W., & Ehsan, M. B. (2016). Crack healing in concrete using various bio influenced self-healing techniques. *Construction and Building Materials, 102*, 349–357. https://doi.org/10.1016/j.conbuildmat.2015.11.006
28. Wiktor, V., & Jonkers, H. M. (2011). Quantification of crack-healing in novel bacteria-based self-healing concrete. *Cement and Concrete Composites, 33*(7), 763–770. https://doi.org/10.1016/J.CEMCONCOMP.2011.03.012
29. Achal, V., & Pan, X. (2014). Influence of calcium sources on Microbially induced calcium carbonate precipitation by Bacillus sp. CR2. *Applied Biochemistry and Biotechnology, 173*(1), 307–317. https://doi.org/10.1007/s12010-014-0842-1
30. Mondal, S., & Ghosh (Dey) A. (2018). Investigation into the optimal bacterial concentration for compressive strength enhancement of microbial concrete. *Construction and Building Materials, 183*, 202–214. https://doi.org/10.1016/j.conbuildmat.2018.06.176
31. De Rooij, M., Van Tittelboom, K., De Belie, N., & Schlangen, E. (Eds.). (2013). *Self-healing phenomena in cement-based materials: State-of-the-art report of RILEM technical committee 221-SHC* (Vol. 11). Springer Science & Business Media.
32. Jonkers, H. M. (2007). Self healing concrete: A biological approach. In *Self healing materials: An alternative approach to 20 centuries of materials science* (pp. 195–204).
33. ASTM C143. (2000). Standard test method for slump of hydraulic-cement concrete.
34. ASTM C873. (2000). Standard test method for compressive strength of concrete cylinders cast in place in cylindrical molds.
35. ASTM C943. (2000). Standard practice for making test cylinders and prisms for determining strength and density of preplaced-aggregate concrete in the laboratory.
36. ASTM C496/C496M. (2000). Standard test method for splitting tensile strength of cylindrical concrete specimens strength and density of preplaced-aggregate concrete in the laboratory.
37. ASTM D790. (2000). Standard test methods for flexural properties of unreinforced and reinforced plastics and electrical insulating materials.

Chapter 8
Urban Drainage Infrastructures Toward a Sustainable Future

Ahmad Ferdowsi and Kourosh Behzadian

8.1 Introduction

Urban drainage infrastructures (UDIs) have had major impact on human and environmental health, urban life quality, and development of cities [1, 2]. UDIs have a long history in urban areas from the time when the traditional open gutters were in place until now that most UDIs' components are underground, and those open gutters/channels have mainly been replaced with closed conduits and piped systems. In fact, the history of using stormwater collection systems coincides with the appearance of human civilization, i.e., thousands of years ago. Although the main goal of drainage systems is to collect surface runoff and flood flows, combined sewer systems in some countries are used to collect and convey both surface runoff and sanitary sewage in the same conduits. As cities grew, the need for larger drainage systems increased, which resulted in more investments in UDIs [3]. However, the performance of combined sewer systems has been found unsatisfactory due to unwanted discharge of untreated wastewater into receiving water bodies that can be the main source of water supply in urban water metabolism [4]. Generally, flooding, erosion, water quality reduction, and environmental issues are probable hazards threatening the performance of UDIs [3].

A. Ferdowsi (✉)
Department of Water Engineering and Hydraulic Structures, Faculty of Civil Engineering, Semnan University, Semnan, Iran

University of Applied Science and Technology, Tehran, Iran
e-mail: ahmad.ferdowsi@semnan.ac.ir

K. Behzadian
School of Computing and Engineering, University of West London, London, UK

Department of Civil, Environmental and Geomatic Engineering, University College London, London, UK
e-mail: kourosh.behzadian@uwl.ac.uk; k.behzadian@ucl.ac.uk

A. Bahrami (ed.), *Sustainable Structures and Buildings*,
https://doi.org/10.1007/978-3-031-46688-5_8

There is no doubt that efficient water infrastructure consisting of many structures and elements [5] can be vital to reach the United Nations' 2030 goals [6], which are known as sustainable development goals (SDGs) and were adopted based on a universal agreement [7]. However, water infrastructure is already under massive pressure from external drivers especially urbanization, population growth, and more importantly climate change.

Despite great importance of UDIs in achieving SDGs, external drivers such as climate change, population growth, and urbanization can undermine the satisfactory performance of UDIs. Although climate change is known as the continuous changes in some climatic variables such as precipitation and temperature [8], the change indicates a rapid rate over the last century [9] due to anthropogenic activities by greenhouse gas emissions from fossil fuels and land-use changes [10, 11]. Temperature, precipitation, relative humidity, and incident solar radiation are some affected parameters in a changing climate. In addition, an increase in world population has been predicted [12], and another projection showed a greater percentage of the world population will be living in urban areas [13]. These stressors can substantially impact natural and human-made structures. UDIs can both affect climate change (e.g., increasing greenhouse gas emissions, acidification, and eutrophication) and also be influenced by climate change (e.g., as a result of changes in urban flooding) [14]. Aging is another problem of current urban infrastructures [15, 16] including UDIs [17] that threatens their sustainability. As displayed in Fig. 8.1, UDIs must overcome a number of obstacles (i.e., aging of infrastructure, climate change, population growth, and urbanization) before reaching a sustainable future, through adaptation strategies. In Fig. 8.2, some problems that happened as a result of malfunctioning of UDIs in Tehran, Iran, are illustrated, which normally threaten the transportation system and public health.

Over the previous decade, various research works studied the impact of climate change [12, 18] and possible adaptation strategies [11, 19]. Reviewing these works demonstrates that the previous studies mainly neglect the role of achieving SDGs and the impact of urbanization; however, most of them emphasize the importance of climate change impacts on UDIs. It should be noted that the performance of current

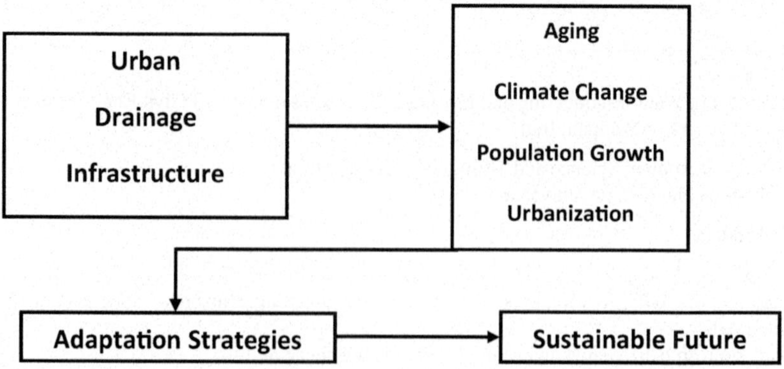

Fig. 8.1 UDI: how to reach a sustainable future

Fig. 8.2 Examples of UDIs' failures in Tehran, Iran

UDIs is affected by climate change and urbanization [3], and applying sustainable drainage can be very challenging for real-world cases [20]. The main aim of the current chapter is to review the impacts of climate change and urbanization on UDIs and potential adaptation strategies to alleviate these negative impacts and help reach a sustainable future.

8.2 Climate Change, Population Growth, and Urbanization

Climate change is one of the most pressing world problems. It refers to the long-term persistent variations in the climate that happen either naturally or as a result of anthropogenic activities [21]. Carbon dioxide, methane, nitrous oxide, water vapor,

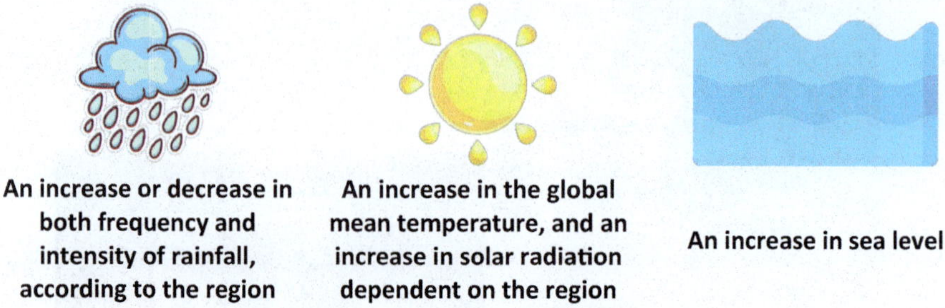

Fig. 8.3 Projected changes of climate variables due to climate change

and fluorinated gases are the most important gasses that are responsible for the greenhouse effect [22]. Human activities are the main reason behind the global increase in greenhouse gases [23] – among all activities, using fossil fuels and land-use changes have had the greatest effect on global carbon dioxide increases [21]. Other impacts of climate change can include changes in migration [24], wildfires [25], extinction of plant and animal species [26], and social and political conflicts [27]. Figure 8.3 depicts some of the featured changes that can happen by climatic changes and affect urban infrastructures.

On the other hand, it has been predicted that the world's population will reach 9.8 billion by 2050 [28]. Another projection, for 2050, indicated that approximately 68% of the entire population of the world will reside in urban areas [13]. Hence, climate change, urbanization, and population growth can be considered the current important world stressors that can jeopardize the conditions of both natural and artificial systems like UDIs.

8.3 SDGs and UDIs

Global goals are an alternative name for SDGs that were set out by the United Nations General Assembly in 2015. Seventeen SDGs with 169 targets are included in the 2030 agenda for sustainable development, which started to be implemented from 1 January 2016. The framework of the goals was prepared in a way that can be acceptable scientifically, politically, and publicly. The final objective of SDGs is to provide sustainable health for all (from planet to local communities), by accounting for poverty, inequality, climate change, environmental degradation, peace, and justice [7, 28]. All 17 SDGs can be categorized into four sections related to people (goals 1–6), prosperity (goals 7–12), the planet (goals 13–15), and peace and partnerships (goals 16 and 17). It can be observed that many of these goals depend on each other, for example, zero hunger in a region affects the poverty of that region. Water is clearly mentioned in goal 6 (clean water and sanitation), which per se underpins many goals such as goals 1 (poverty), 2 (food), 3 (health), 4 (education), 7 (energy), 8 (economics), and 10 (equity) [7].

Since water is harvested, supplied, treated, and delivered through water infrastructure, the dependence of many goals and their associated targets to water infrastructure is undeniable. Flood control is another major task of water infrastructure [29] that is normally handled through drainage (stormwater) systems. UDIs can contribute to reaching goal 1 by reducing climate-related disasters on poor people, goal 3 as death rate and illnesses due to water pollution and contamination can be reduced or eradicated, goal 6 through affordable water production, pollution reduction, etc., goal 9 by making infrastructure and industries efficient, resilient, and eco-friendly, goal 11 by considering the environmental and financial problems associated with water in cities, goal 12 by controlling the release of wastes to water, and goal 13 by increasing public awareness related to climate change impacts and adaptation. Due to the role of UDIs in collecting and supplying water, mitigating flood, conveying water, and contributing in wastewater treatment, they play a significant part in providing safe and affordable water, preserving ecosystems, and other SDGs, which cover seven goals (1, 3, 6, 9, 11, 12, and 13) and their targets.

8.4 Climate Change and Urbanization Impacts on UDIs

In Table 8.1, the main probable impacts of climate change and urbanization on UDIs are reviewed. The capacity of the current UDIs and the quality of water are affected by these drivers. More specifically, floods can be generated because of sea-level rise and increased precipitation that both occur due to climate change. In addition, urbanization as a result of increased desire to live in cities, as mentioned in Table 8.1, increases the risk of flood formation and its associated consequences. Flooding endangers public health, threatens public transportation systems, increases financial losses and number of deaths, and results in untreated water, e.g., wastewater and sewage being released into receiving bodies (sea, lakes, etc.). Based on Table 8.1, the impact of climate change and urbanization can be categorized into four sections in which some problems arise themselves.

Table 8.1 Probable impacts of climate change and urbanization on UDIs

Climate change and urbanization features	Probable impact
A rise in sea level	Higher probability of urban runoff and flooding
Variations in temperature	Changing content of soil water that affects runoff formation
An increase in precipitation	Decreasing water quality; increasing overloading, costs, number of flooded nodes, and water spill from manholes, an increase in sedimentation
Urbanization and population growth	Making urban zones larger and denser, which increases their imperviousness and consequently the rate of floods; land-use changes through removing vegetated areas; a reduction in storage areas

8.5 Performance Improvement of UDIs

Water infrastructure can be divided into three main sub-systems including water supply systems, stormwater systems, and wastewater systems. Needless to say, any new development or rehabilitation of any water infrastructures can be quite expensive, and their construction may take years. Clearly, their failures can result in loss of lives and property damage [30]. Another issue is that many of them were built many years ago which make them more vulnerable.

The impacts of climate change and urbanization on UDIs and their role in gaining SDGs were discussed in the previous sections. The probable impacts of these drivers may cause problems for the operation of UDIs and reaching SDGs. Hence, adapting UDIs to future changes is an urgent need. It was reported that UDIs cannot deal with the effects of climate change and urbanization [1, 12] that necessitate applying the adaptation measures. Design criteria that consider the impacts of urbanization, population growth, and climate change [12] should be added to the future design. Other flood control methods that are listed in Table 8.2 can mitigate flood impacts and reduce the excessive pressure on UDIs due to climate change and urbanization.

8.6 Conclusions

UDIs can contribute to achieving a future sustainable for all; however, it is under pressure from external drivers such as climate change, urbanization, and population growth. The impacts of these stressors not only prevent fulfilling the main functions of UDIs but also undermine reaching SDGs. This chapter investigated the requirements of SDGs in UDIs, the impacts of climate change and urbanization on UDIs, and the adaptation strategies that can be employed to tackle climate change and urbanization and making UDIs ready to achieve the universal goals of the United Nations. The role of UDIs seems to be major for achieving seven SDGs (goals 1, 3,

Table 8.2 Adaptation strategies to counteract future problems in UDIs

Impact	Adaptation strategy
Increased precipitation	Predicted precipitation should be considered for future UDIs; those parts of UDIs that are unable to bear excessive design discharge should be upgraded; changes in precipitation should be considered in simulation models [1]
Urbanization	Employing pervious concrete in urban areas to reduce flood water and to decrease pollution of water [31]; land-use and land-cover modification
Flood control	Using environmentally-friendly solutions (green roofs, vegetation cover, etc.) and source control of water (watershed management); developing prediction models and warning systems; increasing floodplain storage capacity; soil conservation

6, 9, 11, 12, and 13) while climate change and urbanization can cause various problems for UDIs, and the different adaptation strategies were proposed in the literature to mitigate these problems and adapt UDIs to future changes.

References

1. Arnbjerg-Nielsen, K., Willems, P., Olsson, J., et al. (2013). Impacts of climate change on rainfall extremes and urban drainage systems: A review. *Water Science and Technology, 68*(1), 16–28.
2. De Feo, G., Antoniou, G., Fardin, H. F., et al. (2014). The historical development of sewers worldwide. *Sustainability, 6*(6), 3936–3974.
3. Yazdanfar, Z., & Sharma, A. (2015). Urban drainage system planning and design–challenges with climate change and urbanization: A review. *Water Science and Technology, 72*(2), 165–179.
4. Landa-Cansigno, O., Behzadian, K., Davila-Cano, D. I., & Campos, L. C. (2020). Performance assessment of water reuse strategies using integrated framework of urban water metabolism and water-energy-pollution nexus. *Environmental Science and Pollution Research, 27*, 4582–4597.
5. Ferdowsi, A., Valikhan-Anaraki, M., Farzin, S., & Mousavi, S. F. (2022). A new combination approach for optimal design of sedimentation tanks based on hydrodynamic simulation model and machine learning algorithms. *Physics and Chemistry of the Earth, Parts A/B/C, 127*, 103201.
6. Grigg, N. S. (2019). Global water infrastructure: State of the art review. *International Journal of Water Resources Development, 35*(2), 181–205.
7. Morton, S., Pencheon, D., & Squires, N. (2017). Sustainable Development Goals (SDGs), and their implementationA national global framework for health, development and equity needs a systems approach at every level. *British Medical Bulletin, 124*, 1–10.
8. Raju, K. S., & Kumar, D. N. (2018). *Impact of climate change on water resources*. Springer.
9. Dagbegnon, C., Djebou, S., & Singh, V. P. (2016). Impact of climate change on the hydrologic cycle and implications for society. *Environment and Social Psychology, 1*(1). https://doi.org/10.18063/ESP.2016.01.002
10. Kondratev KI, Kondrat'ev KI, Kondratyev KY et al. (2003) Global carbon cycle and climate change. Springer Science & Business Media.
11. IPCC. (2014). Climate change 2014: Impacts, adaptation, and vulnerability. Part A: Global and sectoral aspects. In *Contribution of working group II to the fifth assessment report of the intergovernmental panel on climate change*. Cambridge University Press.
12. Kourtis, I. M., & Tsihrintzis, V. A. (2021). Adaptation of urban drainage networks to climate change: A review. *Science of The Total Environment, 771*, 145431.
13. UNDESAPD (United Nations, Department of Economic and Social Affairs, Population Division). (2019). *World urbanization prospects: The 2018 revision (ST/ESA/SER.A/420)*. United Nations.
14. Behzadian, K., & Kapelan, Z. (2015). Advantages of integrated and sustainability based assessment for metabolism based strategic planning of urban water systems. *Science of the Total Environment, 527*, 220–231.
15. Vahedifard, F., Robinson, J. D., & AghaKouchak, A. (2016). Can protracted drought undermine the structural integrity of California's earthen levees? *Journal of Geotechnical and Geoenvironmental Engineering, 142*(6), 02516001.
16. Roshani, E., & Filion, Y. R. (2015). Water distribution system rehabilitation under climate change mitigation scenarios in Canada. *Journal of Water Resources Planning and Management, 141*(4), 04014066.

17. Francisco, T. H., Menezes, O. V., Guedes, A. L., Maquera, G., Neto, D. C., Longo, O. C., et al. (2022). The Main challenges for improving urban drainage systems from the perspective of Brazilian professionals. *Infrastructures, 8*(1), 5.
18. Willems, P., Arnbjerg-Nielsen, K., Olsson, J., & Nguyen, V. T. V. (2012). Climate change impact assessment on urban rainfall extremes and urban drainage: Methods and shortcomings. *Atmospheric Research, 103*, 106–118.
19. Sørup, H. J. D., Fryd, O., Liu, L., Arnbjerg-Nielsen, K., & Jensen, M. B. (2019). An SDG-based framework for assessing urban stormwater management systems. *Blue-Green Systems, 1*(1), 102–118.
20. Zhou, Q. (2014). A review of sustainable urban drainage systems considering the climate change and urbanization impacts. *Water, 6*(4), 976–992.
21. IPCC. (2007). Climate change 2007: The physical science basis. In S. Solomon, D. Qin, M. Manning, Z. Chen, M. Marquis, K. B. Averyt, M. Tignor, & H. L. Miller (Eds.), *Contribution of working group I to the fourth assessment report of the intergovernmental panel on climate change*. Cambridge University Press.
22. Denchak M (2019) *Greenhouse effect 101*. Natural Resources Defense Council (NRDC). Visited on May 29, 2022. Access link: https://www.nrdc.org/stories/greenhouse-effect-101#:~:text=Earth's%20greenhouse%20gases%20trap%20heat,gases%20(which%20are%20synthetic
23. Lankester, P. (2013). *The impact of climate change on historic interiors*. Doctoral dissertation, University of East Anglia.
24. Feng, Q., Yang, L., Deo, R. C., et al. (2019). Domino effect of climate change over two millennia in ancient China's Hexi corridor. *Nature Sustainability, 2*(10), 957–961.
25. Pausas, J. G., & Keeley, J. E. (2021). Wildfires and global change. *Frontiers in Ecology and the Environment, 19*(7), 387–395.
26. Román-Palacios, C., & Wiens, J. J. (2020). Recent responses to climate change reveal the drivers of species extinction and survival. *Proceedings of the National Academy of Sciences, 117*(8), 4211–4217.
27. Kelley, C. P., Mohtadi, S., Cane, M. A., et al. (2015). Climate change in the Fertile Crescent and implications of the recent Syrian drought. *Proceedings of the National Academy of Sciences, 112*(11), 3241–3246.
28. United Nations. (visited on 5 December 2022). *Take action for the sustainable development goals*. Access link: https://www.un.org/sustainabledevelopment/sustainable-development-goals/
29. Ferdowsi, A., Zolghadr-Asli, B., Mousavi, S. F., & Behzadian, K. (2022). Flood risk management through multi-criteria decision-making: A review. In *Multi-criteria decision analysis* (pp. 43–54). CRC Press.
30. Ferdowsi, A., Nemati, M., & Farzin, S. (2021). Development of dam-break model considering real case studies with asymmetric reservoirs. *Computational Engineering and Physical Modeling, 4*(4), 39–63.
31. Azad, A., Mousavi, S. F., Karami, H., et al. (2020). Properties of metakaolin-based green pervious concrete cured in cold and normal weather conditions. *European Journal of Environmental and Civil Engineering, 26*(6), 2074–2087.

Index

A. Bahrami (ed.), *Sustainable Structures and Buildings*,
https://doi.org/10.1007/978-3-031-46688-5

SPRINGER NATURE

GPSR Compliance

The European Union's (EU) General Product Safety Regulation (GPSR) is a set of rules that requires consumer products to be safe and our obligations to ensure this.

If you have any concerns about our products, you can contact us on ProductSafety@springernature.com

In case Publisher is established outside the EU, the EU authorized representative is:

Springer Nature Customer Service Center GmbH
Europaplatz 3
69115 Heidelberg, Germany

The manufacturer's authorised representative in the EU is Springer Nature Customer Service Centre GmbH, Europaplatz 3, 69115 Heidelberg, Germany. If you have any concerns regarding our products, please contact ProductSafety@springernature.com

Printed and bound by CPI Group (UK) Ltd, Croydon, CR0 4YY
01/10/2026
02225883-0003